The Fluoride Question

Anne-Lise Gotzsche

DAVIS-POYNTER
LONDON

First published in 1975 by
Davis-Poynter Limited
20 Garrick Street London WC2E 9BJ

ISBN 0 7067 0165 8

Printed in Great Britain by
Bristol Typesetting Company Limited
Barton Manor St Philips
Bristol

Contents

Introduction

In September 1973 I wrote a feature for *The Sunday Times* on fluoridation which caused the Fluoridation Society Ltd to launch a nationwide attack on me personally and inspired the British Dental Association to accuse *The Sunday Times* of having done 'enormous harm to the cause of dental health' as well as giving in to an 'anti-fluoridationist coup'.

In this attack, British fluoride promoters used much the same language and methods as American ones, and the nature of such attacks show how difficult it can be for a journalist to publish research which happens not to coincide with the line taken by medical and scientific authorities. Such attacks – and the pressure which is sometimes put on editors to publish only pro-fluoride copy – also demonstrate the extent to which fluoridation is a political and not a purely scientific or dental controversy. A great many interests are at stake, and some have very little to do with dentistry.

It seems to me that dentists are taking it upon themselves to decide, not only in matters of public dental health but also in matters of environmental health, a very much more complex subject of which they have little knowledge. In so far as environmental health concerns industry and the waste products produced by industry, as well as the legislation which is necessary to cope with such pollution, the dental profession could be said to be trespassing into territory which belongs to other experts. In the sheer political fervour with which they try to force fluoridation on the community, they could even be said to be usurping the power of Parliament. As far as fluoride is concerned, politicians today are taking advice from *dentists* on matters which lie far outside the dental field

– and the dentists in turn often pretend – or believe – that these other problems do not exist.

They are doing this at a time when figures compiled from recent official US statistics show that, after over 25 years of fluoridation, some of the longest fluoridated cities in North America, Grand Rapids, Newburgh, and Evanston, have approximately *twice* as many dentists per unit population as the average figure for the whole country, and almost twice as many as the average number for the 'naturally fluoridated' towns.

There are many reasons for dentist-population ratios, but these figures do not point to a dramatic decrease in dental decay, nor to dentists being done out of their jobs by fluoride, as they are fond of claiming. Quite the contrary. America is going through a remarkable dental crisis with dental decay affecting 95 percent of schoolchildren and with 25 million Americans toothess, despite all their fluoride.

Yet in Britain, the British Dental Association now suggests that pressure should be put on the education authorities to include pro-fluoride projects in school syllabuses and dentists are constantly encouraged to play politics with fluoride, not just dental politics but public politics.

The real importance of this issue is not so much fluoride itself as the fact that this is the first scientific controversy to be put to the vote across the world, and to be subjected to all the usual, well-financed publicity stunts of ordinary political electioneering.

And yet the leading players – the scientific fluoride proponents – are not elected by the public, do not have to answer to the public for their mistakes, and are responsible to no one but each other for their decisions. Policy is decided behind closed doors and no respectable opposition to the official party line is tolerated.

Dentistry is not the only sphere of science where a few men may hold great power and influence unknown to the general public, though often only too painfully well-known to their own rank and file.

A science-based society would be ill-advised not to have a hard and close look at current scientific discontent. The very fact that even the poor, ill-organized, and amateurish lay opposition to fluoridation has been so successful for so long, shows that at least the general public have a kind of understanding of what is going on, and that even lay people now seem to sense that the scientist who deals in ideas may be just as dishonest and unfair in trying to sell his ideas as the manufacturer and the salesman.

There is another point, a point made by the British Dental Association. In attacks on me they have complained angrily that they had not been 'consulted'. They had of course been consulted, but that is of no consequence. No political journalist is under any obligation to 'consult' with the powers-that-be before he writes about them, nor is he stopped from questioning their actions or pointing out their mistakes. There is no earthly reason why a science journalist, or anyone else for that matter, should 'consult' the British Dental Association before writing about dentistry.

The sheer arrogance of dentists who want newspaper editors to suppress letters from people who don't agree with them is perhaps what this whole quarrel is really about.

Science is full of quirks, fads and fashions, lobbying for grants and position, personal feuds, and genuine disagreements. The bill, as always, is footed by the taxpayer, the electorate, the consumer. And yet too many scientific societies still cling to the archaic belief that they have inherited some sort of superior and moral right not to be questioned. When, as in this case, academic interests happen to coincide with those of politicians and of industry, there is trouble indeed. Thus Shropshire County Council could, in the autumn of 1973 come out and propose adding fluorosilic acid to the drinking water at no less than 22 remote and unguarded points without anyone except a retired police and factory surgeon making a fuss – apart from the 'vociferous minority' or 'cranky' anti-fluoridationists.

Today good and reasonable dentists state privately that 'fluoride helps a little but not very much'. The statistics would seem to bear out this verdict. Unfortunately good and reasonable dentists prefer to stay in the background and just get on with their work. They are rarely to be found among the fluoride gospel preachers, and the public rarely hears about them. How cautious we should be about those scientists and academics who do decide to play politics – without following the normal democratic rules of the political game – has been splendidly illustrated in a now almost forgotten dental incident.

Back in 1945 two now famous and distinguished American dental researchers, Michael G. Buonocore and Basil G. Bibby, warmly recommended the use of *lead* fluoride in dental prevention, claiming that lead was far more effective than fluoride and lead fluoride far more effective than any other types of fluoride. They did this in an excellently reasoned research paper which is now quite a collector's item – though it causes medical researchers to throw up their hands in horror.

That our children are not now given lead tablets at school to prevent tooth decay is perhaps a reflection of the fact that the public may have more sense than they are sometimes given credit for.

ANNE-LISE GOTZSCHE

'Fluoridation of water supplies would be marvellously beneficial to young children and, in the absence of proof of toxic effects of such a measure, it is difficult to understand the ethics of those who deny its benefits to them. It amounts, in my opinion, to a *conspiracy of wickedness* against young children.'

Surgeon Rear Admiral W.I.N. Forrest,
Guy's Hospital Gazette, 24 February 1973.

I

Yousef Mustapha's Problem

'We have endemic fluorosis in some sections of the country because of the large intake of tea . . .'

'In case we do not have fluoride in our water, I will try, using the information I have from the American Dental Association, to convince the government to start a project of fluoridation . . .'

These two remarks come from Yousef Mustapha Sehni, described as Jordan's Father of Modern Dentistry, in an article in the *Journal of the American Dental Association* in January 1972.

It is a fair example of what the fluoridation controversy is all about, and of the kind of contradictions which fluoride promoters will sometimes accept without batting an eyelid. Jordan, it seems, has endemic fluorosis because of the high fluoride content in tea and the remarkably large number of cups of tea drunk by some people in that country. Yet – to the dental mind – it is desirable to add still more fluoride to the drinking water. How this could result in anything but yet more fluorosis does not appear to have worried the American Dental Association on this occasion.

What is really worrying to the rest of us is how apparently a whole generation of dentists, doctors, and politicians could have accepted such arguments without finding anything odd about them at all. And it should be remembered that the basic arguments for and against fluoridation have hardly changed since the war, even though new research and new doubts have been encouraged by the environmental movement. It would be sad if the clamour from the ecology lobby should obscure the fact that what went wrong with fluoridation, went wrong right at the start, and almost everything that has gone wrong since, has gone wrong because it was based on the same original premise.

Society may no longer be in such eager search for wonder drugs and cure-alls as it was in the 40s and 50s, but the underlying emotions and the normal twists and turns of scientific controversy are not likely to change. In this respect it is fair to say that fluoridation is basically an American story – even today American researchers are from time to time accused (by their British counterparts) of looking for miracle drugs with an almost religious fervour. This happened recently over methadone, the synthetic replacement for heroin. In fairness too, however, fluoridation should also be seen in its historical context: the idea was born not so long after penicillin, but a long, long time before thalidomide. If the early promoters were over-enthusiastic, then at least some of them were so with the very best of intentions.

All the same, fluoridation is a lesson in something which may very well happen again, and there may be more urgent issues than this one which it will not be so easy to pass judgement on in hindsight, and which may also not be quite so funny – for fluoridation is, in a perverse way, *funny*. For those who take a historical view it could well go down as a monumental bad joke, a classic of the century for generations of schoolboys to come. When you scrutinize the annals of this controversy it is indeed often difficult to know whether to laugh or cry.

The pros make bland and sweeping statements about 'safety', the antis never run out of stories which sound as if they were straight out of 'Scoop': research subjects are 're-shuffled', figures misquoted, high-fluoride areas recorded as low and low ones as high. If the teeth in an unfluoridated control town are doing well, then the unfluoridated area is quite illogically fluoridated as well. The beneficial effect of fluoride is 'proved' in an area which – it turns out later – didn't actually get the fluoridated water through some fault in the water system. Fluoridation promoters move into areas with public campaigns, without first checking the level in the local water, and discover to their chagrin afterwards that the water already contains the 'optimum' amount – despite the bad teeth. Points and commas get misplaced in the text and 1.2 ppm becomes 12 ppm, and for years it is proudly claimed that 12 ppm is 'absolutely safe'. Unerupted teeth are counted as 'sound teeth' for statistical purposes. The fluoridators forget to tell the public and the press that areas with high fluoride levels are ripe with dental fluorosis and malocclusion. It comes as a surprise to sceptics to discover how often these claims are true.

Some may be genuine mistakes. Some certainly are not. What is clear is the fact that most of the scientists who have taken a close look at the issue and who have come out against, have some very harsh criticisms of the statistical evidence as well as the scientific methods employed. What is also certain is that such scientific criticism has often been refused publication in the proper dental and medical journals, especially those journals which would reach the dental community – though a great deal has been published elsewhere. Thus the kidney researcher and the toxicologist and the orthopaedic specialist may worry about fluoride, but the dentist and quite often the ordinary doctor remain blithely ignorant. Not only the public, but these doctors and dentists themselves are ill-served by such public and editorial attitudes.

People tend to see fluoridation as a straightforward case of being for or against. This is not the case. There are roughly five groups involved in the battle.

At one extreme are the dentists – a fairly tight international clique of leading names who meet up frequently at international symposia and conferences and who have staked their careers and reputations on the safety and benefit of fluoride, sometimes in conflict with their own better judgement and occasionally as a result of financial inducement.

In the English-speaking world dentists as a whole are in favour, though this is not the case in, for instance, Sweden. A few wellknown names in dentistry have spoken out against, but on the whole even those who might have done so in a more liberal climate, tend to feel that it is not 'worth their while' to risk getting into trouble with their professional organizations. The active promoters crop up again and again on just about every and any committee that has ever recommended fluoridation and it is often a case of the same individuals wearing many different hats. They make sweeping public statements such as 'fluoride is essential to life', 'fluoride is absolutely safe', 'fluoride cuts dental decay by half' (or by 60 percent or 70 percent or 90 percent, depending on the passion of the speaker). They can be easily identified by the sheer, undaunted sweep of their statements and the total certainty with which they make their proclamations. It should be said that some, if not all, are excellent and distinguished researchers in the many other aspects of their subjects and they would undoubtedly have reached the top of their profession, even if fluoride had not existed.

Behind these scientists – but not necessarily connected with them – is an anonymous group of industrialists who do not wish the public to become aware of the enormous problem of coping with fluoride pollution and fluoride waste disposal, as well as those who manufacture fluoride toothpastes. These are also wholeheartedly in favour of fluoridation, though it is not quite clear why the toothpaste manufacturers should be.

At the other extreme is the 'vociferous minority' of lay anti-groups who feud amongst themselves, are sometimes extremely emotional about it all, quote and misquote everybody *ad nauseam*, who send photocopies of personal letters round the world to perfect strangers (very useful to journalists, that), who reprint without permission, and who have a deplorable tendency to use sentences like 'death in a tube of toothpaste', 'dental mafia', and 'government-sponsored hoax'.

They have, according to a friend of mine in the British consumer movement, 'poisoned the media' with their extravagant claims – though this equally extravagant remark could just as well be applied to the fluoride promoters who, after all, have more money to spend on the media.

Behind this group are a growing number of scientists who sometimes publish, but more often keep quiet about, their doubts and reservations, who in some cases appear to have been subjected to considerable intimidation, but who often are equally terrified of being quoted by the lay antis. Some have spoken out boldly and survived, though often at a price. Thus John B. Polya, Professor of Chemistry at the University of Tasmania, has retained his position by continuing 'research activities in fields unconnected with this issue', and so has Dr. P. R. N. Sutton, of the Faculty of Dental Science at the University of Melbourne – though when his book *Fluoridation, Errors and Omissions in Experimental Trials* (Melbourne University Press) appeared in 1959, the type was mysteriously destroyed by an unknown person and £400 had to be spent on re-setting it for the second edition. Yet in America, Dr Ludwick Gross, chief of cancer research at the Veterans Administration Hospital in Bronx, NY, has got away with describing fluoride as an 'insidious poison' and apparently avoided trouble.

They have their cranky hangers-on, of course. In the November 1962 issue of the *Journal of the American Dental Association*, a few anti-fluoridation scientists were lumped together with people who were obviously cranks and 'health food faddists' in what was equally obviously an attempt to discredit these scientists. Most of the leading names in the field were not mentioned, and dentists reading that issue would have come away with the idea that no really serious criticism of the issue existed. In fact the long list of people who have asked uncomfortable questions about fluoride now includes Professor Barry Commoner. It would be fair to say that fluoridation is an issue which has caused doubts in the very best and most cautious minds.

Lastly, there is among the extremists a small group of people who sell fluoride tablets or wish to add fluoride to milk, and who are therefore not in favour of adding the stuff to the water. These people are sitting very comfortably on the fence, trying to have their scientific cake and eat it and making soothing statements such as 'there is bigotry on both sides' – which is certainly fair comment.

The issue these groups are quarrelling over is not purely the simple one of adding fluoride to the water. There are three main lines of argument and acrimony: the actual dose of fluoride consumed, the question of mottled teeth and caries resistance, and the politics arising out of claims that scientific methods have been fraudulent.

First, the dose. This is currently the most fashionable, partly because of the new interest in ecology and in nutrition. It is known now that not only may the food content outstrip the amount recommended for adding to the drinking water (and many healthy foods have a high natural content), but in industrial and polluted areas the food content alone may already be too high. Barry Commoner is one of the scientists who has tried repeatedly to get the US Public Health Service to give him data on the total dose, without much success.

Superficially, the dose can be worked out fairly easily. It is usually accepted that children should have no more than 1 mg a day, adults no more than 2 mg. More will not necessarily cause any harm, but dental fluorosis has occurred at lower levels. Natural or traditional diets rarely exceed 1 mg a day. Industrial diets do. Adding another 1 mg per litre of drinking water (i.e. 1 ppm) therefore already seems a bit of slightly dubious brinkmanship, especially considering that processed foods made in a fluoridated area add to the total load and may double or treble the fluoride content measured in ppm (or parts per million).

Moreover, children would get another 1 mg from each fluoride tablet, plus yet another 1 mg a day from gobbling up fluoride toothpaste. In Graz, Austria, fluoride tablets have recently been withdrawn from schools for this reason and, in Anglesey, the Medical Officer of Health has recommended that fluoride toothpastes are not stocked by the local chemists (despite being an ardently pro-fluoridation man). Thus the thin line between benefit and danger is an extremely narrow one.

Even if you confined yourself to water only, the 'optimum' or safe dose for adults would be only around four pints, or more precisely, two litres. Moreover, women who suffer from nonspecific urethritis or bladder complaints, and who are sometimes advised to drink half a pint of water every twenty minutes, would reach their 'optimum' dose in exactly two hours 40 minutes.

Water apart, it has been shown by John Marier, of the Division of Biology, National Research Council of Canada, that the total dose in the West in industrial areas may now be 5 mg a day, and in Japan as much as 11 mg a day has been recorded. This latter dose is similar to the dose which has – on at least one occasion – caused trouble when administered as a drug under medical supervision.

It is known that diet may play an important role in the way the body handles fluoride. Calcium and vitamin C have a mitigating effect on the action of fluoride, and it is thought that inorganic iron and fats may enhance it. Calcium in this respect is very important, since water is usually hard in those areas where fluoride occurs naturally, while soft water tends to have a low fluoride level, around one tenth of that recommended for fluoridation. There are exceptions, obviously, but these are also frequently troublesome areas with regard to fluorosis.

One may therefore ask: why add another milligram per litre of water, if some people are already getting too much? And how can you calculate the effects of a pharmacologically-active, dose-dependent substance, if you do not know what the total dose is? The authorities have been repeatedly accused of ignoring these two points – though it would be wrong to say that they have completely neglected them.

The WHO's report on Fluorides and Human Health in 1970, warns that people living in a fluoridated area should take care to avoid foods high in fluorides. In 1963 the Department (then, Ministry) of Health warned local medical officers in Britain that people in high-fluoride areas should never leave a kettle on the boil and should always throw away water used for cooking vegetables. In 1953 the Report of the United Kingdom Mission to the US which recommended fluoridation, suggested that 'the use of products which are naturally high in fluoride content, such as bonemeal tablets, or of lozenges, dentifrices, or chewing gum, to which fluoride has been added, should be avoided where the drinking water has been fluoridated'.

Yet one has the impression, in Britain, that not only the Department of Health but the British Dental Association endorse not just fluoridation, but fluoride in a sort of blanket fashion: fluoride is good for you, any fluoride, anywhere, in any amount, except excessive amounts – and it's not easy to see what these learned authorities regard as 'excessive'. In the final analysis it boils down to individual reactions, individual diet, individual state of health, but this cannot be admitted by those who advocate that everybody should have *more*, regardless.

Secondly, the controversy over mottled teeth which has been with us for much longer. The term 'mottled enamel' appears first to have been used in medical literature in 1915, but the 'dark spots' were commented on as early as 1771. One writer, L. Meredith, stated in 1878 that:

'White, yellow, or brown spots of various sizes and irregular shapes may exist on the outer surfaces of the teeth. . . . The incisors are the only teeth attacked in the great majority of cases, but occasionally others are also. Sometimes several of the teeth in one or both jaws may be so severely affected as to scarcely look like teeth, appearing as if they had been badly eaten and discoloured by some corrosive agent.'

The link with fluoride is not new, nor is the observation that such teeth may also display an absence of caries. This was established scientifically by G. V. Black and F. McKay in 1916. The idea of adding fluoride to prevent caries was also first aired in the 19th century. The physiological effects have been known for decades, at least from the clinical point of view.

At first there will be an initial hardening of the enamel and resistance to decay, a delay in the occurrence of cavities, and a delay in the eruption of the teeth. The shape and size of the teeth may also be affected. The most consistent pattern of fluoridation statistics is a delay in the onset of decay by about one year, and the difference in decay between fluoridated and unfluoridated teeth varies between half a tooth to two teeth per mouth. In Britain, according to *Fluoridation Studies in the UK and the Results Achieved after Eleven Years* (HMSO 1969), the difference was 0.9 of a tooth by the age of 14 and 0.8 at the age of 8. If one allows for the delay in the onset of cavities of just over one year, the development of dental decay was roughly similar in both groups. This benefit is thus a temporary one, and often a very small one too.

If excessive fluoride consumption is continued, there will be mottling of the teeth – also called dental fluorosis – and in time the teeth may become stained. Later such teeth may become brittle and eventually it may be impossible to make them hold ordinary fillings. Two well-known researchers, M. C. Smith and H. V. Smith, at the University of Arizona Agricultural Experimental Station, observed years ago that 'when decay does set in, the result is often disastrous'. According to some writers, such mottling is now common in some fluoridated communities in the US. Dental journals even advertise adhesives to cover up the stains.

Traditionally mottled teeth in America are called 'Texas teeth' or 'Colorado stain', and in Italy, in volcanic areas in and around Naples, they are called *'denti neri'* or *'denti scritti'*. This phenomenon occurs in animals too, for instance in cattle. It could be argued that in so far as fluoride is 'natural' – as the pros claim – mottling certainly is a 'natural' problem in certain types of environment, such as a volcanic one.

Mottling is not purely a cosmetic effect, but the earliest sign of what, with very high fluoride ingestion, could become crippling skeletal fluorosis – though skeletal fluorosis (and other complaints) may occur in the absence of mottling, if the excessive fluoride intake begins after the teeth have been formed. Skeletal fluorosis is extremely unlikely to occur in the normal course of events in temperate climates, but is well-documented in hot climates. Sometimes it is caused by excessive tea drinking rather than by high water levels.

This then, put very briefly, is what the scientific side of the quarrel is about. And this is what all the passion is about.

Yet it is not, according to F. J. McClure, chief of the Laboratory of Chemistry, National Institute of Dental Research, US, and a leading fluoride promoter, 'a subject for public debate by informed scientists'. And those who try to claim otherwise, have been described by Dr Harold Hillenbrand, former executive director of the American Dental Association, as 'scientific primitives' and 'emotional primitives'. Dr Miles R. Markley, of Denver, Colorado, is on record as saying that people who oppose fluoridation 'have been proved to be psychopathic'. And Dr Donald Galagan, assistant chief of the Division of Dental Public Health, the Public Health Service, Washington, has described them as 'hate-mongers'. Professor G. Neil Jenkins, of Newcastle University, one of Britain's leading fluoride promoters, claims that 'The agitation against fluoridation has been largely based on emotion rather than science'. But when sentences such as 'when people discover that they don't die like rats . . .' crop up regularly in scientific literature, one may wonder who is most emotional. At times, it seems, people have been on the brink of actual physical fighting.

Devonport, a town in Tasmania, was made to fluoridate against the will of its citizens in 1971. According to one Devonport councillor, Mrs Noreen Batchelor, and a local JP, M. T. W. Davenport, a point was reached where the municipal council tried to lock the state officials out of the waterworks, while the state government in turn threatened each councillor with an immediate fine of 100 dollars plus a further fine of 40 dollars a day, unless they agreed to do as they were told. Eventually they did.

In Britain, methods are a little more subtle; in America this is not always so. Ralph Nader has alleged that the US Public Health Service has a file on all the major critics of fluoridation with extraneous personal information which is then released strategically to various sources and media throughout the country. He has also claimed that:

'The US Public Health Service has a totally paranoiac mind on this subject. And it's very serious because if it is paranoid on this issue, what is it going to be on other issues?'

I myself, when I first started researching this subject, was warned that 'the boys down in London' would 'fleece' me. I have heard a leading British fluoride promoter described as a 'nutcase' by his own colleagues, and even such words as 'hoax' and 'plot' originate in the medical and dental establishment, not among the antis.

This, obviously, is language any layman can understand. There has rarely been a scientific or medical issue followed with such intense interest and emotion by the lay community – and with such laxity by the scientific establishment. The latter is all the more surprising, considering that fluorine is one of the more exciting and certainly difficult elements and one which greatly excited scientists in the past. Yet today such regular discussion as one would normally expect on a subject such as this in medical journals seems to have been almost totally stifled by the need to conform to the dogma that fluoride is absolutely safe.

How did it happen? I am not sure I have an answer. Unless it is in the element of mystique which fluoride undoubtedly has, the intellectual somersaults which were necessary originally in order to regard as safe and beneficial what was formerly regarded as dangerous, the faith which was necessary to stick to the new view – and, on the side of the antis, the danger, excitement, and moral virtue of fighting compulsive human consumption of 'rat poison'.

My own first concern is not with whether the pros or the antis are absolutely right or wrong on any particular point. My concern is with what J. Bronowski calls the 'monstrosity of certainty', the fact that science – like life – can never be too certain, and those who try and claim that fluoride, or anything else in science, is absolutely certain, or 'absolutely safe' are betraying the rest of us, and science as well – apart from being arrogantly stupid.

How arrogantly stupid can best be illustrated with my favourite anti-fluoridation quote, from the Nov.-Dec. issue of *National Fluoridation News* (US):

'Mason City, Iowa, . . . was the scene of an interesting fluoridation campaign . . . The fluoride drum beaters descended upon Mason City with the usual misleading literature. The hometown paper was enlisted in the crusade. Endorsements by experts were introduced. Civic groups joined up with the parade. Local doctors issued statements about the deplorable state of the children's teeth. The decay rate was so appalling that the local dentists couldn't handle the terrible situation. Dr Charles Henshaw, employee of the Iowa Department of Health, came to town to display his charts and graphs to show how 1 ppm would lower the dental decay of the Mason City children's teeth by 65 percent.

'Then someone tested the Mason City water supply and found that it already contained 1.25 ppm – a little bit more than the magic amount of fluoride. Almost the same thing happened at Ottawa, Illinois, only the water there contained 1.3 ppm of natural fluoride.'

One must hope that Yousef Mustapha doesn't make the same mistake in Jordan.

II

'A Plot Against Fluoridation'

'It may have been disappointing that they found that both sides are about equally informed, interested and aware of the fluoridation issue; that liberals are no more likely to favour fluoridation than conservatives; that freedom of choice was of little or only secondary importance; and that anti-intellectualism or anti-scientism were no more common among those who voted against than among those who voted for fluoridation.'

These words came from John W. Knutson, a leading fluoride promoter from the School of Dentistry at the University of California's Los Angeles Center for the Health Sciences. It was reported in the October 6 1970 issue of the *British Dental Journal* and supported by Gordon M. Williams, chairman of the BDA Dental Health Committee, with the words:

'As soon as dentists recognize their responsibility in the politics of fluoridation, their performance will be outstanding. In politics, the emphasis is on propagandizing rather than on educating. In politics, the emphasis must be on commitment rather than detached objectivity. . . . In other words, a dentist does not need to know all the vast scientific background to fluoridation – all he needs is the knowledge that fluoridation is safe, effective and practical, and enough enthusiasm to convince other people that this is so.'

This fairly and squarely puts the standpoint of the British Dental Association. The dental community clearly regards fluoridation as a political issue and not a scientific one. In America you can get a Ph.D. on the study of the lay anti groups. A lot of money has been spent doing surveys into the social, financial, political, and psychiatric backgrounds of 'positive' people (who vote *for* fluoridation) and those who vote against. The politics of the thing even extends into chemistry and basic scientific terminology.

In Europe fluoride is called 'fluor', but in the English-speaking world there is still argument about whether it should be called fluor*ide* or fluor*ine* in analogy with chlorine, iodine, and bromine, the other three members of the so-called 'halogen' group of elements.

The early fluoride promoters disliked the word fluor*ine* because they feared that lay people might think that they were proposing to put the elemental toxic fluorine gas into water, which of course is not the case. The WHO, like many chemists, still prefers fluor*ine* as the more correct. The *n* helps to make clear exactly what you are talking about: thus if a patient is given 1mg of sodium fluor*ide*, approximately 45 *percent* will be *fluorine (as fluoride)*, the rest will be sodium. Unfortunately many writers are extremely vague on this point, and it can sometimes be difficult to work out whether they mean the dose to be twice as high or perhaps only half. Confusion over the dose has occurred even in medical journals.

This problem was the subject of argument in a legal fluoridation battle in the High Court of Dublin and the Supreme Court of Ireland in 1963 and 1964. The antis stuck angrily to the *n* and insisted on talking about fluorination. The court decided in favour of the *d*. That disputes concerning basic chemistry can actually be settled in a court of law may amuse scientists, but there the matter rests. Scientific writers usually like the chemical symbol F^- to describe the fluoride (-ine?) ion, but otherwise the word fluor*ide* is now normally used to describe the fluorine part of the compound (and I shall use it myself in this book), to distinguish it from more volatile and dangerous forms of elemental fluorine.

If all this sounds complicated, it is still only a small part of the arguments and counter-arguments which have been thrashed out by local councils all over the world for the past 20-30 years, to the point where one may wonder whether a degree in chemistry should not be made a condition of being elected to local government. Even so, the real cause of the passion it arouses can only be understood in a historical context.

In the 1930s H. Trendley Dean of the US Public Health Service was engaged in investigating the effects of fluoride occurring naturally and noted a connection between mottled teeth and an absence of caries. As already mentioned, this was not new, but Dean was the person who really put this observation on the map in a big way. He is now popularly called the Father of Fluoridation. Dean was certainly far from being as extreme in his attitude as some of his followers, and he also called for further investigations into the long-term effect on health. In 1939, however, a biochemist called Gerald J. Cox, of the Mellon Institute, had come up with the idea of fluoridating the water supply (although this again was nothing new. Apparently Rudolf Steiner thought about it in the 1920s and others before him). Cox was supported by a lawyer, Oscar Ewing, and both had been associated with the Aluminium Company of America. The antis make much of the fact – or claim – that both had an interest in diverting attention from the fluoride waste products produced by that company. Ewing, a former counsel to the company, later became Director of Social Security of the US Public Health Service and played an important part in persuading that organization to endorse fluoridation.

Originally, however, the US Public Health Service was unwilling to recommend fluoridation.

According to James W. Benfield, of the Columbia School of Dental and Oral Surgery and an anti-fluoridation man, Dr Daniel Ziskin, chief of the diagnostic department of the same school and one of the official advisers on the issue was pretty scathing about what he regarded as the unscientific methods proposed for the planned fluoridation experiments. Ziskin predicted the controversy that followed, though he did not live to see his prediction come true.

In 1945 two experiments were started, one in Grand

Rapids, Michigan with Muskegon as control town (although Muskegon itself was later also fluoridated), and one in Newburgh, NY, with Kingston as control area. Ziskin's comments on the Newburgh experiment was that it was inconceivable to him that the New York State Department of Health would consider subjecting 40,000 people to the addition of one of the most toxic substances known to mankind even in minute quantities without ever having done an animal experiment to prove its safety. Moreover, he complained that there were no provisions for proper dental examinations with X-rays, no provisions for the study of the possible effect on adults, and moreover no provision for double-blind measures, such as are normally used in medical trials. Even today, newcomers among the scientific objectors are still harking back to this experiment in particular, echoing Ziskin's remarks, because Newburgh, along with the other 'classic' American fluoridation studies, is still used as 'proof' of benefit as well as harmlessness by health authorities all over the world.

The Newburgh and Grand Rapids trials were supposed to last ten years. However, there were people who were not willing to wait.

As early as 1943 the fluoridation idea had been given impetus by a story in *Reader's Digest* entitled 'The Town Without a Toothache' – the town in question being Hereford in Deaf Smith County, Texas – despite objections by the local dentist, a Dr Heard, who complained that Hereford had an abundance of the many other minerals which are beneficial to dental health not only in the water, but in the vegetables, milk, grains, and meat from animals raised locally. He wrote a pamphlet about this in 1951, which, of course, was far too late. The bandwagon had started, and already two years earlier, in 1941, Wisconsin cities had accepted fluoridation mainly as a result of the persuasive efforts of the State dental society and two of its members, Dr Francis A. Bull, of the State Board of Health, and Dr. John Frisch, a Madison dentist. Frish and Bull launched a remarkably intensive nationwide effort to obtain PHS endorsement of fluoridation before the ten-year period had lapsed. The strength of their emotional involvement, and of their scientific objectivity, may be illustrated by a story which is going round the anti-fluoridation circuit, that Frisch even went so far as to label his ordinary tap from the *un*fluoridated water supply 'Poison'. Opposing biochemists, he is quoted as having said, 'didn't know a fluorosed tooth from a bed-pan' and he referred to their experimental findings on rats as 'hogwash from the biochemistry department'. He has been described as 'a man possessed'. 'Fluoridation became practically a religion with him', according to Geoffrey Dobbs, a senior lecturer in forest botany at the University College of North Wales, Bangor, and a well-known British anti-fluoridation man. These quotes originate in *The Fight for Fluoridation* by D. R. McNeil (1957) – an American fluoride promoter.

On a more sober level, Michael Wollan, a research associate in Law, Science and Technology at the National Law Center of the George Washington University, published a study of what followed in a report in the July 1968 issue of The George Washington Law Review. (Wollan died in a car accident at the time this report came out).

According to Wollan, Frisch and Bull criss-crossed the country attempting to persuade dentists and community leaders of the benefits of fluoridation. This campaign was so intensive that Herbert Bain, an official of the American Dental Association, 'came to dread the meetings because those guys would show up and never let you off the hook'.

By 1949 Frisch and Bull were beginning to reap results. Privately they told the Public Health Service that 'they had better get on the fluoridation bandwagon if they did not want their image as leaders in public health to be permanently tarnished'. The climax of the mounting pressure came in late May, 1950 at the annual meeting of state and territorial dental health officers in Washington. Dr Bull 'once again buttonholed every major Public

Health Service official attending the conference. In particular he concentrated on Dr Bruce Forsyth, assistant surgeon general and chief dental officer for the PHS. Bull drew Forsyth aside and told him that he was 'being made a sap out of' because 'before long the PHS would be the only major health organization refusing to endorse fluoridation'.

At a meeting at this time Dean argued that since the New York and Michigan studies were only half completed, it was still not time for the PHS to take a position. He stuck to this view, but the others won. On June 1, 1950 the PHS announced its endorsement of fluoridation. Since then most other major health and medical agencies, including the WHO, have been won over with similar tactics, many simply taking their lead from the PHS.

Eighteen months later evidence was presented at an investigation by the House Select Committee to investigate the use of chemicals in food products, which indicated that a number of questions had not been thoroughly explored. For instance – as already mentioned – the committee concerned itself that while physical examinations were performed on children in the Michigan and New York projects in order to detect possible adverse effects, no such examinations were performed on adults. A pattern had already been set for later examinations of this sort where 'those with chronic illness and disease known to affect bone structure were excluded', according to John Marier, of the National Research Council of Canada.

Dr John Knutson, of the PHS also admitted that studies had never specifically gone into the question of the possible effect of the addition of artificial fluorides to water on children who are suffering from malnutrition. Yet, the February 1952 issue of the *Journal of the American Dental Association* warned that 'low levels of fluoride ingestion . . . may not be safe for mal-nourished infants and children because of disturbance in calcium metabolism'.

Dr Isador Zipkin of the Public Health Service admitted

that toxicity studies on rats were not completed until at least a year after the PHS formally endorsed fluoridation. When congressman A. L. Miller asked a representative of the National Cancer Institutes if any experiments had been carried out as to ill effects in pregnant women and in people with chronic diseases, a PHS official said he knew of none, and so did Dean. It was thus clear already that fluoridation was endorsed before there were adequate grounds for claiming that fluoridation was safe, and that was the verdict of the Select Committee. Nobody, it seems, took any notice of the verdict – fluoridation rolled on across the land as if nothing had happened and no questions had been asked.

At this point the argument took on an almost farcical air. In 1951 a meeting was held by the state dental directors with the PHS and the Children's Bureau. The purpose was not to examine the pros and cons, but to discuss promotion, and the state dental directors were under the impression that the minutes of the meeting were not to be published. They spoke freely, especially Dr Bull. The following remarks were made:

'What do we care what happens to rats? You know these research people . . . they can't get over their feeling that you have to have test tube and animal research before you start applying it to human beings.'

'There have been several instances where groups have promoted fluoridation of the local water supply only to find that the supply already contained the optimum amount.'

'Now why should we do a pre-fluoridation survey? Is it to find out if fluoridation works? No. We have told the public it works, so we can't go back on that.'

'Water contains a number of substances that are undesirable, and fluorides are just one of them.'

'Don't use the word "artificial". Don't let them raise the question of rat poison if you can help it. And certainly don't use the word "experimental". Fluoridation is not an "experiment" but a "demonstration".'

'The medical audience is the easiest audience in the world to present this thing to.' (Implying that doctors are naïve.)

'When they (the antis) take us at our own words, they make the most awful liars out of us.'

This last remark – and indeed the whole meeting – has gone down as an anti-fluoridation classic. The minutes were kept secret for seventeen years. Then a few copies were published, got into the hands of a doctor who is campaigning against fluoridation, and distributed by him.

Since then official attitudes have hardly changed. In the early 1960s the Pennsylvania Department of Health published a leaflet with advice about how to promote fluoridation, how to avoid questions on toxicity, how to ridicule its opponents, and how to expose their personal backgrounds. One famous paragraph – which has become another anti-fluoridation classic – gives advice on how to drag God and religion into the discussion:

'God has given us the element of fluoride so that we can use it for our own benefit. Now He has shown us the best way it can be used to protect our children's teeth against tooth decay. He has inspired our scientists with the idea that the easiest and best way that our children, and our children's children, can get it is by putting it in the water He has given us. We are blessed to have discovered that tooth decay can be prevented by the simple and safe procedure of drinking it in our water. Water is our most important resource. And now God has endowed us with the knowledge that by using this important resource we can almost miraculously prevent two thirds of tooth decay in our children.'

At the same time some fluoride promoters seem to have become blasé about their public remarks. According to Gunnar Bergström, a Swedish dentist then working as an associate professor of dentistry at Texas University, professor J. C. Muhler of the University of Indiana said in 1963 that the great benefit of fluoridation to *dentists* was the fact that the enamel became so brittle that dentists needn't waste time on ordinary fillings, but could concentrate on the more profitable work of fitting crowns.

At Texas University when Bergström was there, the medical faculty felt so pressed that they decided that the word fluoride must not be mentioned in the common rooms for five years – at that time that university had been taking a stubborn anti-fluoridation stand.

Wollan says that the PHS 'has adopted an aggressively defensive attitude towards critics of fluoridation which tends to discourage meaningful exchanges of views. For instance, Public Health Service scientists have refused to publicly discuss or debate with scientists who are critical of fluoridation' and 'the ability of the PHS to offer a balanced assessment is limited by the strong commitment to fluoridation it has voiced since 1950. Moreover, the PHS attitude has occasionally led it to take action fostering a charged atmosphere in which objective evaluation of fluoridation is difficult.'

For example, when the American Society for Fluoride Research held its first meeting in September 1966, several well-known anti-fluoridationists were openly associated with the group. Although presentations were made by pro-fluoridationists, anti-fluoridationists and scientists not committed to either view, the PHS feared that the conference might receive publicity which would influence an impending referendum in Detroit on fluoridation. The PHS therefore successfully urged the American Dental Association headquarters in Chicago and Detroit to release a statement charging that the ASFR 'is only a sounding board for fluoride opponents'. The *Detroit News* picked up the release and ran a front page story with the headline, "Research Talk Called Plot Against Fluoridation".

The PHS took similar steps in the autumn of 1967 with respect to an organization called the International Society for the Study of Nutrition and Vital Substances, a European-based group of scientists that passed a resolution opposing fluoridation.

The US Public Health Service, learning of this, hastily

issued a statement claiming that American anti-fluoridationists had infiltrated an international society, pushed through a resolution condemning fluoridation, and were now using it for their own purposes in the US. Later the PHS learned that the American committee was formed long after the resolution had been passed and that none of the committee members had anything to do with the drafting or passage of the resolution, and the PHS therefore issued a revised statement.

'Nevertheless,' says Wollan, 'this case and the case of the ASFR indicate that the strength of its commitment to fluoridation can prompt the PHS to take hasty action that compromises its ability to provide a detached, comprehensive assessment of comments and research on fluoridation.'

Ralph Nader puts it this way, commenting on criticisms of fluoridation:

'Oh, these are interesting questions, except that nobody is asking them in this country and still getting public subsidies for research.'

It isn't only in the US, however, that repressive tactics have been used. A symposium was held in Bern, Switzerland, in October 1962 dealing with the potentially poisonous aspects of fluoridation, with the title, 'Is Fluoridation Dangerous?'. A book was to be published of the papers presented at this symposium. According to Professor Albert Schatz of Temple University – the co-discoverer of streptomycin, the drug that cures tuberculosis, a leading researcher in dentistry as well as cancer and originally a soil microbiologist, who claims that he lost his share in the Nobel Prize for his part in discovering streptomycin because of his outspokenness against fluoridation and other dental controversies – the following happened:

'. . . neither the symposium nor the book received any support, financial or otherwise, from dental organizations. On the contrary, some groups actively opposed both the meeting and publication of the report. Consequently the two were carried out independently of the dental profession . . .'

'The symposium was originally scheduled to be held in Holland, but because of opposition from dental interests in that country it was transferred to Bern. There, others subsequently opposed publication of the book because it, like the symposium, would inevitably raise questions about the safety and wisdom of fluoridation (and about some who had initiated this programme and were continuing it).'

Apparently two years elapsed, according to Schatz, between the symposium and publication, because further efforts were made to suppress the book. For instance, one medical and dental publishing house, it seems, invested some 10,000 Swiss francs in setting up the text. But was then warned that if it went ahead and published the book, 'the dental community would stop patronising it'. The publisher then dropped the book. It appeared in print only because the editor, Dr T. Gordonoff, Emeritus Professor of Pharmacology at the University of Bern and a leading researcher in goitre and the thyroid, persisted with his efforts to get the book into print. It was finally published by Schwabe & Co.

That, at least, is how the story goes. Schatz himself is not only outspoken but has a gift for words and 'quotes' which journalists find difficult to resist. His verdict on the fluoridation controversy in America reads as follows:

'Ever since US dentistry "created" fluoridation it has been forced to defend it in the face of increasing worldwide opposition from many responsible scientists. . . . As a result the reputation of US dentistry has become irrevocably bound to the fate of fluoridation. A stage has now been reached where the rejection of fluoridation will irreparably discredit the American Dental Association and the National Institutes of Dental Research of the US Public Health Service.'

Professor Polya in Tasmania is equally scathing:

'Never before has the need to justify administrative decisions affected the design and trustworthiness of supposedly scientific work. Never before has compulsion anticipated scientific evidence, and never before have political authorities rejected the advice of so many distinguished scientists in favour of amateurish claims.'

Ludwick Gross puts it this way:

'The plain fact that fluorine is an insidious poison, harmful, toxic, and cumulative in its effects, even when ingested in minimal amounts, will remain unchanged no matter how many times it will be repeated in print that fluoridation is "safe".'

It would be fair to point out here – with no reflection on Dr Gross – that even scientists have from time to time been confused about exactly when fluorine, or fluoride, should be regarded as truly toxic. However, there can be no doubt that unreasonable statements about 'safety' have been repeated in print in a most unreasonable manner, and the repetition does not, indeed, alter the basic chemistry of the issue. What causes otherwise reasonable people to make such statements can be difficult to comprehend. In a few cases there have been suggestions by the antis of commercial or financial involvement.

On the whole one should not, perhaps, attach too much importance to these isolated individual cases. Scientists are usually in science for the love of it and rarely primarily for financial gain and corruption, in so far as there is any, is more likely to be of an elusive intellectual kind rather than overtly financial.

That a scientific community as a whole may be a victim of outside commercial pressure is quite another matter. And certainly there has been direct commercial interest in promoting fluoridation through grants and subsidies. James W. Benfield, of the Columbia School of Dental and Oral Surgery, thus remarks on the interest of the Sugar Research Foundation which already in 1949 aired its interest in finding out 'how tooth decay may be controlled effectively without restriction of sugar intake'. That, of course, should surprise no one. Benfield also mentions that fluoridation was endorsed by the American Water Works Association 'despite the fact that the majority of water engineers then, as now, oppose the program'. A more explicit accusation has come from Dr

Willard E. Edwards, formerly a US Navy corrosion control engineer for the Pacific Area, and now working in Honolulu, namely that he himself 'severed my membership with them (The National Association of Corrosion Engineers) when I found they are largely sustained by chemical and aluminium companies which sell or have fluoride as a by-product of their industries'.

Only in Germany has there been any outright and strong national opposition from water engineers, but there is no doubt a lot of tacit and quiet doubt about its benefits. A water engineer in one of Britain's major and recently fluoridated cities told me on the phone, 'we are not used to being given an impartial view in this matter from the British medical profession', and recently the South East of Scotland Water Board has – according to their critics – 'become a law unto themselves' and defied the decision of the Edinburgh Corporation to fluoridate the area.

In America the Aluminium Company of Canada actually at one point in the 50s published a full-page advertisement recommending 'Alcan sodium fluoride', though the ads were withdrawn when the antis began using them as proof of commercial involvement.

More recently, an almost hilarious commercial angle was reported in *New Scientist* in March 1971: the American Chemway Corporation was said to have discovered that fluoride never actually penetrates the teeth because both are negatively charged. To overcome this they suggested that the enamel charge be reversed by means of introducing into the saliva a fluoride salt to form an electrolyte and then pass an electric current through the mouth during toothbrushing, thus turning the mouth into a kind of organic battery. The company, though, also warns of physical discomfort due to sore fillings and headaches as a result of this treatment.

American television fluoride commercials are equally entertaining.

One commercial, sponsored by the Atlanta-based Northern District Dental Society and reported in the

31 July 1972 issue of the *American Dental Association News,* goes like this:

'Destry Fluorides Again!'

The plot: Flossie Molar, the heroine, is about to have her teeth attacked by Dirty Don Decay. Masked-puppet-man Destry Fluoride answers Flossie's last-minute pleas for help and saves her from Dirty Don with a blast of fluoride treatment. Having saved another damsel, Destry rides into the distance shouting, "Hi-yo, Smi-leeeeee".'

More difficult and more serious are accusations that professional associations such as the American Medical Association may be influenced by commercial interests – though as late as this year, in February 1973, the Association was alleged by a writer in *Science* to have dissolved its Council on Drugs 'as a sop to the pharmaceutical industry. For a number of years the AMA, which derives a large portion of its income for its annual budget from advertising of drugs, has been a captive of and beholden to the pharmaceutical industry'. The anti-fluoridationists suspect that AMA may be open to influences from other sides as well.

In October 1944, the *Journal of the American Medical Association* published an editorial which asserted:

'We do know that the use of drinking water containing as little as 1.2 to 3 parts per million of fluorine will cause such developmental disturbances in bones as osteosclerosis, spondylosis and osteopetrosis, as well as goiter, and we cannot afford to run the risk of producing such serious systemic disturbances in applying what is at present a doubtful procedure intended to prevent development of dental disfigurements among children.'

It is extremely doubtful that they would repeat this statement today.

The World Health Organization has come under fire too. It has been alleged from several sides that on the day when the WHO originally voted on fluoridation, only 60 out of 1,000 delegates were left in the hall, and these happened to be those who were in favour. At the World Health Assembly of the WHO in the US in July 1969 a resolution was passed which recommended fluoridation where fluoride intake 'from water *and other sources* . . . is below optimal levels'. (My italics). The resolution also calls for more research into, among other things, the fluoride content of diets, the effects of fluoride on the human body, and the effects of excessive intake from natural sources. This, however, does not appear to have made the slightest difference to the WHO's official attitude to the problem. At a press conference which I attended at the British Medical Association headquarters in London in 1972, the WHO press officer warmly recommended fluoridation as being 'absolutely safe'. Their 1970 Report *Fluorides and Human Health* is a singularly contradictory and selective document – impressive as it may look at first sight to the un-medical local county councillor. No scientist who had spoken out against fluoridation was invited to contribute, and the editor, professor Yngve Ericsson, is known the world over for his strong pro-fluoride bias.

The WHO appears in a general way to be extremely vague about water levels, and has been criticized by scientists for its lack of specifity even when considering totally different substances. Thus Dr Eric Hamilton, of the Institute for Marine Environmental Research, Plymouth, wrote in the 20 September 1972 issue of the *Journal of the Royal College of General Practitioners*:

'Referring to WHO "permissible levels" it was often difficult to see how they had been arrived at and very often very little work had been done.' One may safely assume that WHO endorsement of fluoridation doesn't mean a thing, scientifically. Like everybody, they seem merely to have jumped on a bandwagon.

It would be naïve, of course, to assume that there is such a thing as an international 'dental mafia' – but the list of strong and outspoken statements hinting at intimidation is such that one could fill a small book with these quotes alone. The lay antis may be forgiven if they sincerely believe that this affair is a case of a dental Watergate. Thus the following statement from W. B.

Hartsfield, Mayor of Atlanta, Georgia in early 1961, has been frequently quoted.

'The general public does not realize the gigantic power structure which is pushing fluoridation. The PHS spends over 840 million Dollars per year in grants, cooperative programs and salary supplementation. Through this means they have welded the State and County organizations into a completely docile and responsive organization, amenable to the slightest pressure from Washington.

'Endorsements by many national organizations are obtained by the same persons who are active simultaneously in all of them. Public Health officials endorse fluoridation, then put on another hat and call themselves a State organization, endorsing it all over again.

'The World Health Organization is financed by the US Public Health Service; of course they endorse fluoridation.

'This power apparatus is spending millions every year in a mad effort to force all waterworks to accept their recommendations. USPHS has dozens of public relations men: every State organization has one; all are industriously putting out prepared stories endorsing fluoridation.

'No school, college, or independent medical research institution dares to be critical of fluoridation because they receive Public Health grants. Likewise, no big food, beverage or drug company will dare speak critically of fluoride because they are under the supervision of the Food and Drug Administration, a branch of the PHS.

'One big brewery official told me that their own research 'indicated grave questions about fluoridation, but they dare not speak out'.

If this sounds drastic, it is beginning to look as if fluoridation is not the only field where scientists in America are feeling repressed or intimidated. Indeed, at the time of writing, one research group is busy collecting evidence of such accusations from fellow researchers. They seem to have had a particularly good response in the biomedical sciences, and feel that these people 'would undoubtedly agree with' one of Albert Schatz's more outrageous, but ever quotable, remarks about the same old faces meeting up year after year in different parts of the world reinforcing their own and each other's prejudices. I have come across a similar expression of apprehension and fear of speaking out or drawing attention to yourself, from one Pennsylvania scientist travelling through London who was doing some controversial but interesting and worthwhile work in cancer. If the press, he thought, made too big a noise about it, he might lose his grant.

It should be said, though, that when Americans do speak out, they do so in a grander and more courageous fashion than the British. The American anti-fluoridation literature reverberates with the remarks and statements made by such people. America is a very strange country, and a few more quotes should help to describe what is going on just below the surface.

'In Antigo, Dr Dorzeski, Health Commissioner, was represented as an ardent promoter of fluoridation. But at that very moment he wrote to Dr Albert Schatz that he was in truth opposed to fluoridation.'

'A. C. Baumann, D.D.S., of Cleveland, Ohio, asked to be permitted to present scientific evidence against fluoridation to approximately 1000 Cleveland dentists. He was informed that his material would have to be submitted for approval first to the fluoridation committee of the society and then to the council before it could be presented to the society at large. He was not allowed to speak.'

'In a letter by J. E. Waters, D.D.S., he stated that he had prepared a series of articles against fluoridation for publication in the Californian Journal *Compass*, and that he was told that he would be automatically expelled from the American Dental Association if these articles appeared under his name.'

In 1964 the Michigan House of Representatives decided that more research was needed before recommending fluoridation, and in their report stated, among other things:

'The commercial aspects of fluoride should also be studied as well as methods employed in promoting fluoride among physicians, dentists and scientific organizations. Reports were submitted of a request by 85 doctors to their medical society to further examine this subject. These were ignored.'

The Michigan Representatives also found it odd that 170 communities in the US who adopted fluoridation, should have discontinued it.

The *National Fluoridation News* regularly carries 'testimonials' from erring dentists. In the Nov-Dec 1971 issue, Dr Casimir R. Sheft, a graduate of the University of Maryland Dental School, and later Chief Surgeon with one of the US Navy operating bases, filled one and a half pages of print with his objections to fluoridation. In the Jan-Feb issue the same year, a similarly long list of objections came from Dr U. L. Monteleone, a practising dentist in Allentown whose eyes were apparently opened to the truth when he went down to Puerto Rico and discovered the appalling state of the children's (fluoridated) teeth. Dr Monteleone was later involved in a celebrated anti-fluoridation scandal when he was sacked from the Allentown Hospital allegedly because of his opposition to fluoridation.

I have not tried to double-check or verify the particular cases just mentioned. In any case, there seems to be so many of them, that it would be an impossible task for any one person to undertake. What is important, I think, is the fact that the endorsement of fluoridation by official professional organizations and government agencies, does not by any means signify that the members of these organizations also agree, nor that the issue has been settled and is no longer 'a matter for public debate'.

I would like to think myself that this whole awful affair could be blamed on the enthusiasm, eagerness, and good intentions of the early promoters, at least in the case of some of them. It is very difficult.

Even the *Journal of the American Dental Association* seems to have had a bout of conscience about over-enthusiasm – though in a slightly different context.

In the July 1972 issue they published an extraordinary editorial in which the Journal attacked what it called 'dental evangelism' and the way in which many younger dentists claim to have invented 'preventive dentistry' while the older ones thought that this was what they had been practising all along.

'The dental evangelist is becoming less and less believable, and he is sounding more and more like a huckster . . .'

Huckster, perhaps, is a nice American word for a very American phenomenon.

In reply to all this – and it has been stated and spoken and printed many, many times before – what do the British health authorities have to say?

On 19 October 1972, Sir George Godber, Chief Medical Officer at the Department of Health (until the end of 1973), stated:

'The baseless opposition to this humane and safe measure in Britain makes a mockery of one aspect of the care of children's health.' He asserted that 'the drive to promote fluoridation of water supplies was still blocked by a vocal minority on false and unscientific grounds', and added, 'it may be that some future finding will show that there is a further additive which, if it were used in soft water supplies, would reduce the incidence of coronary thrombosis'.

This last remark is one which would inflame the antis, and which sounds ominously familiar to anyone who has delved into the murky waters of this controversy.

It is alleged that Mendès-France once suggested adding antabuse to French drinking water to cure the French of alcoholism. And in 1968 Melvin M. Ketchel, of Tuft's University School of Medicine, suggested adding 'fertility control agents' – presumably some sort of contraceptive drugs – to the water as a way of controlling the population explosion. These, he said, should lend themselves 'to being easily and *unobtrusively* included in the intake of everyone'. (My italics). Further drugs should then be given to people to make them fertile again in those cases where it was desirable that they should have children.

Lithium is another candidate as a way of combating mental illness, assuming that it would work without also adding tranquillizers as well (which is not the case at present).

One British psychiatrist suggested to me in a letter that it should be added to the diet of British prisoners:

'Then we could set about clearing the prisons of these impulsive psychopaths . . . .'

The step towards adding it to the water is not a long one, and that has in fact been suggested. A doctor from the Cincinnati General Hospital, Dr Barry Blackwell, wrote in the *American Heart Journal* in January 1972:

'By adding *Lithium fluoride* to our water supplies, can we anticipate growing old with teeth, hearts and sanity assured? Will we have learned enough about the pitfalls of evaluating such claims to distinguish science from science fiction?'

III

A Statistician's Joy

'. . . it has never been claimed by the Department (of Health) that dental decay is caused by an absence of fluoride.'

This interesting statement appears in a 16-page circular sent out to Medical Officers following criticisms of the official HMSO 1969 report recommending fluoridation in Britain. The circular was aimed particularly at the National Pure Water Association and seems to indicate that the Department will, on occasion, take a lot of trouble to prepare replies to supposed 'cranks' and 'fanatics', even allowing for a ministerial tendency towards wordiness.

The sentence also appears in a singularly muddled paragraph which ends with the statement that 'there are many examples in medicine of preventive measures which, although unrelated to the actual cause of the disease, have an important effect in diminishing its prevalence.'

It is difficult to see what the Department may have been thinking of. *Preventive* measures are usually preventive exactly because they are related to the disease in question, unlike such drugs or methods as may have been designed to check it, reverse it, or simply relieve the pain. All the same, this is a good example of the official thinking behind the fluoridation policy and can perhaps be regarded as the basis on which that policy is recommended. It should be born in mind when examining the actual statistics.

Fluoridation is a statistician's joy. Despite the fact that most fluoridation statistics are difficult to interpret because of the consistent lack of information about the total and actual dose consumed from all sources, because of the lack of information about important factors like diet, breast-feeding, oral hygiene, hardness of the water, the other minerals such as strontium and vanadium which are known to be beneficial, and despite the constant refusal to apply proper age correction to allow for the initial delay in the onset of caries and the eruption of the teeth – despite all this, fluoridation statistics are important.

The importance lies in the way in which they show how easy it is to play around with figures in order to prove whatever you may wish to prove, at least to your own satisfaction, if not to that of your opponents.

A word of kindness may be necessary. It should be remembered that even in the mainstream of medicine, population surveys can be difficult to assess and even at the best of times, give rise to frequent and heated controversies. A laboratory test can usually be repeated by others without too much ado. A population survey is a major effort and one which tends to be compared not with a replica of itself, but with other major surveys in different areas where other factors may play a part, and perhaps where different methods may be necessary. Moreover, because of the sheer amount of work and time involved, the factors surveyed have to be limited, and other factors – perhaps unknown at the time or not thought important at the time – may later alter the picture. All such surveys should therefore be treated with due reserve. What is proved – if anything is proved – is never a *causal* relationship, only a possible association. This has become blatantly obvious in heart disease research where some quite ludicrous ideas such as keeping budgerigars or not going to church or committing adultery, have been linked with the disease apparently in all seriousness.

One could therefore easily think of 101 reasons why fluoridation statistics should be totally meaningless. That they are not meaningless is shown by the fact that there is a fairly consistent overall pattern on a world-wide basis: that of an initial delay in the onset of caries, of roughly one year. Even in those cases where a survey shows the teeth in the fluoridated town to get worse, the teeth in the control town will get worse at the same rate, only slightly ahead, with the same time-lapse phenomenon.

Indeed, that there *is* an initial cavity-delaying effect has never, to my knowledge, been denied by anyone. This association can be found in the graphs and curves of averaged-out statistics, but does not hold with individual mouths, or individual age-groups, or individual teeth. In fact, the individual variations may be so stark that only a dentist would accept them – and I'm not being bitchy saying this. When it comes to measuring dental decay, dentists are unfortunate in having a number of fundamental problems.

A major one is the difficulty in defining exactly what a cavity is, and how to count it. Just counting cavities won't work because two cavities may grow and merge to become one, and then you end up having one cavity but more decay. I doubt whether this would happen all that often – attacks on the teeth tend to be single attacks, but the problem is there.

Another obstacle is the fact that there may be disagreement about exactly when a cavity is a cavity. Ideally – it is commonly thought – you need X-rays for a proper diagnosis. Ordinary clinical assessment with the naked eye may be unreliable. Different dentists may arrive at different counts. But even X-ray examinations may not always solve the puzzle.

Dr F. B. Exner, a well-known American antifluoridationist, wrote in the June 1967 issue of the *National Health Federation Bulletin* of one examination of 833 students where 1,662 cavities were found by clinical examination and 1,372 by X-ray – but only 237 were the *same* cavities. Decay, he concludes, can't be measured.

Yet another point of confusion – when related to fluoride intake – is the fact that the permanent teeth even in the normal course of events erupt at different times, which means that different teeth would have received the 'benefit' of fluoride for different lengths of time at any one particular age or date. Even without fluoride, children tend to show great variation in decay rate at different ages. Add to this the delaying effects of fluoride – whether or not you acknowledge its presence – and you are in trouble with your figures. At least you will be in trouble with your conclusions.

To this should be added yet more problems. For instance, it is a common observation that girls tend to have more caries than boys, though the cause has not been established. What *has* been established – by Professor D. Jackson at the University of Leeds – is that there is a strong genetic factor playing a part in tooth decay, producing similar patterns in the same types of teeth in different populations. This observation may not upset the fluoride theory, but it does seriously challenge the nearly 100-years-old idea that cavities are caused by bacteria in acid fermentation: the teeth in a mouth share roughly the same bacterial environment, but only some teeth decay. In animals too there is – in the laboratory at least – a strong species-difference with regard to the most likely sites of attack.

All this is the stuff that science is made of, and the usual thing is for people to have it out in the professional journals and eventually reach some kind of working hypothesis or compromise. This works well enough up to a point, provided you remember that it is a compromise, when you come to draw your conclusions from your experimental findings. Dentists have a crude, but workable, compromise, the best perhaps that could be found, and the trouble only starts when they make sweeping statements at the end of the day.

An interesting example is Hartlepool.

In 1971 a survey was published in the *British Dental Journal* by J. J. Murray of the Institute of Dental Surgery at the Eastman Dental Hospital in London. This compared the teeth of adults over the age of 15 in Hartlepool which has a natural fluoride content of 1.9 ppm, with York, a low fluoride area. The results showed the number of edentulous or toothless persons over the age of 35 in both towns to be roughly similar. As the normal decayed-missing-filled (DMF) count is meaningless for people over the age of 34 because many people have their teeth extracted for reasons other than caries, the author gives individual figures for each group. These, oddly, show that there was more decay in high-fluoride Hartlepool, though there were more missing and especially more filled teeth in York. The author concludes from this that he has proved that the high fluoride level in Hartlepool has a *'lifelong beneficial effect'*. One could equally well have concluded that people in York go to the dentist more often to have their teeth filled. The author, moreover, seems to have arrived at his 'proof' by averaging out the three different counts and finding a 40-50 percent reduction, though he has just explained that it is statistically meaningless to do so.

No matter. Whenever British fluoride promoters today claim that 'the effect lasts for a lifetime', they are thinking of this particular survey, remembering with pride the conclusion, but not the figures.

In 1970 another survey was conducted in Hartlepool, among 100 female factory workers. This showed that as much as 85 percent had some degree of dental fluorosis, and 25 percent had severe fluorosis 'leading to self-conscious reaction'. One girl had been so distressed that she had part of her teeth replaced with dentures and another 13 were using 'harshly abrasive toothpaste' in an attempt to whiten the teeth. Would it be too wicked to conclude from both surveys that, yes, the *dental fluorosis* lasts a lifetime, though there is no improvement in the caries situation?

As already mentioned, the usual statistical way of counting decay is to count DMFs – decayed, missing, and filled. Two other common measures is to count the number (or percentage) of mouths with ten or more decayed teeth, and the number (or percentage) free of decay. These three methods are all right in themselves, though it is difficult to compare them with each other when dealing with the same population. Moreover, the cautious observer should be particularly careful about not being fooled by DMF percentages. Fifty percent may sound grand, but one tooth is 50 percent of two, just as eight are 50 percent of sixteen.

That is not quite the way it is worked out of course. The 'protective effect' (i.e. the difference) is expressed as a percentage of the caries experience of the 'unprotected' control group. That way one may arrive at some flashy-sounding percentages.

It is of interest to laymen to note that using this method of 'percentage reduction' to compare the effects before and after fluoridation in the *same* area, *a difference of one tooth* may come out as a *percentage* of 16, 25, or 29.4 depending on the actual number of teeth involved in the calculation. The lower the number of teeth, the higher the percentage. One rotten tooth is a higher percentage of two rotten teeth than it would be of six rotten teeth. This is what happened in the 1969 HMSO Report in Britain with children in the seven, ten, and eleven age groups.

The result is that the 'benefit' in terms of percentage reduction in cavities declines sharply with age, startingly so. The famous Table 7 of the British report begins with an impressive figure of 68.5 percent (given as 67) reduction in the three-year-olds after fluoridation (and a real difference of just under two and a half teeth). By the age of eight the reduction has gone down to 43 percent, by the age of ten to 29 percent (given as 31), and by the age of 14 to 18 percent (given as 19). From the age of eight onwards the real difference hovers round the one-tooth-mark, regardless of whether you do a before-and-after calculation or compare the fluoridated areas with the unfluoridated ones.

This is a straightforward lesson in how perfectly legitimate scientific methods should be treated with caution and accepted only in their proper context and setting. The percentage reduction method is used in science exactly because it enables you to show up small differences in a big way. It is a useful method, but to employ it for political slogans, to stamp the results on envelopes and to splash them across posters is an unethical and reckless way of misleading the lay public.

This is apart from the fact – as indicated above – that the authors in this case got some of their own percentage calculations wrong. Though the mistakes are minor ones, it is wise in this game to check the details (even when dealing with toothpastes: a 50 g tube of paste may contain 'only' 0.4 percent of a fluoride compound, but it works out at 200 mg.)

The only way to judge the real state of affairs is to know the actual number of decayed teeth. This, of course, is also done. There is a tendency among fluoride promoters to present the figures in cavities per 100 mouths (especially in the US), while the antis prefer single mouths. The reason is obvious. Ninety cavities in 100 mouths sound like a lot more than 0.9 of a decayed tooth in a single mouth. As ultimately we are all concerned with the state of the teeth in each individual child, it is a good idea to be a simpleton and insist on knowing how many teeth will be decayed in each single mouth. In Britain – as cannot be repeated too often – the Department of Health's own report shows the difference between fluoridated and unfluoridated teeth to be 0.9 of a tooth per mouth by adolescence, and slightly more or slightly less in earlier age groups. And *that* is what the Department of Health and the British Dental Association claim as 'Cuts Dental Decay by Half!'.

This routine is one which all antis are well versed in, and it is almost moving to see how totally different observers in different parts of the world, quite independently, all seem to have arrived at the same conclusions and made the same discoveries. Anyone who seriously undertakes to look at the small print and individual figures in fluoridation statistics, almost invariably finds himself plodding down the same short road to the amazing truth. And more often than not he will soon find himself smacked like the naughty child who pointed fingers at the emperor.

This whole incredible muddle over methods, percentages, and conclusions has been with us from the beginning, starting with the early classic fluoridation 'demonstrations' in America. These have been seriously criticized by Philip R. N. Sutton, at the Department of Oral Medicine and Surgery, University of Melbourne, who got a second version of his book out in 1960 (despite the attempt to destroy the type). Sutton, among other things, recommends that DMFs should be counted not per mouth, but per 100 *erupted* teeth. Unless this is done 'comparisons between the rates prevalent in the test and the control cities, and those seen in different years, are not valid'. As a matter of fact, the delaying effect shows up perfectly clearly in the conventional method and graphs drawn from the two sets of figures will coincide if allowance is made for that delay. However, Sutton's method would of course be a better way of investigating this phenomenon.

There is a strong consensus among scientists critical of fluoridation that fluoridation has not been investigated the way it should and that the methods used as well as the conclusions drawn are scientifically unacceptable. One is sometimes tempted to agree with the antis that the only thing that *can* be counted with any certainty is the number of times fluoridation is declared to be beneficial in the text accompanying the figures.

This has been done – tongue-in-cheek – by the incorrigible Albert Schatz from Temple University and published in the *Pakistan Dental Review* – a journal which, along with the *New York State Dental Journal*, appears to be a receptacle for all such papers and articles which do not honour current dental gospels.

Schatz analysed the 1969 HMSO report on fluoridation in the UK and came up with the following statistical results:

'The official report repeats in one way and another no less than 89 times that fluoride protects against caries.

This is an average of 10.7 times per page, and 1.8 times per paragraph. It uses the word 'benefit' as a verb, noun, and adjective, 30 times. That comes to an average of 3.6 times per page and 0.6 times per paragraph.

'The preface of the report has seven paragraphs and these tell us 13 times that fluoridation protects against caries. That is an average of 1.9 times per paragraph. The preface also extols the alleged benefits of fluoridation an average of once in every four of its 52 lines.'

Good clean fun for learned professors, even though one hesitates to draw the conclusion that the equally learned authors of the report have deliberately tried to mislead. But one needs a lot of generosity to believe in their honourable intentions.

A typical example in this report is the Anglesey study. Gwalchmai was the study town, fluoridated in 1955/56. Bodafon was the control town – though fluoridated in 1964, three years before the end of the study which was 1967. Then Holyhead was lumped together with these two as a 'study' town – although the text points out ten times that Holyhead had no control. Holyhead also received mixed water, from both fluoridated and unfluoridated sources and – remarkably – had vastly better teeth than the other two. What that *proves* is anyone's guess.

The final results between Gwalchmai and Bodafon for the 8-10-year-olds in 1967 show half a tooth's difference between the two, and for the 11-14 age group in 1965, 1.4 teeth.

Kilmarnock provides similar fun. Kilmarnock was fluoridated in 1956, and stopped fluoridating its water in October 1962 because the local authorities didn't feel it had made much difference. Some children's teeth improved remarkably, but other's deteriorated.

Five percent of the children who were 5 years old in 1961 had more than 10 decayed teeth at that point, but by the time they were 7 years old and still fluoridated, as much as 10 percent had 10 or more decayed teeth. Conversely, the percentage of children free of decay among the 5-year-olds in 1956 was only 6 percent, but during fluoridation it rose to 20 percent, so these children obviously benefited. However, when fluoridation was stopped, the children who were 7 years old in 1968 showed a percentage of only 8 percent with more than 10 teeth decayed, compared with 23 percent among those who were 7 years old during fluoridation in 1961. Thus there was more decay at this age level during fluoridation than after. Kilmarnock, incidentally, also had very slightly better teeth to start with, and the teeth in the unfluoridated control town Ayr improved in some age groups at a steady rate during the same period.

Add to this the mathematical confusion when fluoride promoters claim simultaneously that you shouldn't stop fluoridation because then the teeth will at once deteriorate, and that the effect 'lasts for a lifetime', and it seems that people could go on arguing forever. This, indeed, they do in Anglesey where (at the time of writing) six out of eight local councils want to stop fluoridation but have been refused permission to do so by the County Council.

To properly understand fluoridation one must keep in mind not only the figures for mottling, but also the fact that quite often a fluoridation campaign is accompanied by a general campaign in dental health, nutrition, and oral hygiene. This seems to have happened in Birmingham.

Birmingham was fluoridated in December 1964. In the Birmingham Health Report, 1968, it was reported that extractions had gone down from around 6,000 to around 2,500, but at the same time fillings had gone up from 2,500 to 10,000. These figures are for the group that received the 'full benefit' from fluoridation, the under-fives. Both trends started *before* the city was fluoridated. Nevertheless the Medical Officer of Health concludes that fluoridation 'has played a substantial part' in reducing the number of extractions and increasing the number that were 'fillable'. Simpletons like me would suspect that what happened was simply that more people took their children to have their teeth repaired before it was too late.

The report also mentions that 166 children under five were supplied with full upper or full lower dentures and that altogether 673 dentures were supplied to children in this age group. (Presumably the remaining 507 dentures were partial dentures.) One could hardly base sweeping claims for the benefit of fluoride on such evidence.

It is important to remember in this maze of figures that the so-called and proverbial 'ordinary' people – some of whom may be greater simpletons than me – have tended to assume that 'Cuts Dental Decay by Half' means that half their children's teeth that would have decayed would be saved. One can only wonder at what they think when their own children still end up with false teeth, as unable to chew and bite into tough, healthy food as granny. Or perhaps they think that if the children *only* have upper *or* lower dentures, then half the teeth were indeed saved?

A word about the official classification of mottled or hypocalcified teeth should be said here.

Fluoride promoters are fond of talking about the 'strikingly good appearance' of mottled teeth (unless or before they stain), 'the most beautiful looking teeth that anyone ever had' and in America dentists like to write about the snow-white 'mother of pearl' look of such teeth – though healthy teeth ought really to have a very slightly 'yellow' hue like ivory despite toothpaste manufacturers' claims to the contrary. They must be odd dentists who do not know this. It should be remembered too that there may be other kinds of 'mottling'. Mottling may occur in low-fluoride areas, though the cause is not altogether clear. Children born to mothers who take tetracycline drugs during pregnancy will also display effects similar to those caused by fluoride. Different types may not be clearly distinguished clinically, but are under the microscope.

The original classification of fluoride mottling and the

degree to which it may or may not be desirable was hazy enough to provide ammunition for both pro and anti guns to this day, and is still accepted by dental authorities as a basic rule for what is or isn't acceptable. Laymen may be surprised at what is acceptable.

The guide rules were laid down by Dean: Normal, Questionable, Very Mild, Mild, Moderate, Moderately Severe, and Severe. Moderately Severe was officially dropped in 1939, but is still sometimes used.

'Mild' is explained as 'the white opaque areas in the enamel of the teeth involve at least half the tooth surface. . . . Light brown stains are sometimes apparent generally on the superior incisors'.

Dean found 'mild' fluorosis even at 0.4 ppm.

In Professor G. Neil Jenkins' book, *The Physiology of the Mouth* (Blackwell Scientific Publications, 1970), 'moderate' is defined thus:

'With intakes of 2 ppm about 10 percent of the teeth were graded as showing "moderate" fluorosis, in which the whole enamel surface showed either the opaqueness or the brown stain. Fewer than 25 percent of the teeth were free from some defect.'

On the whole, though, they are cautious about linking specific degrees of fluorosis with water levels, or perhaps vague might be a better word. They are equally vague, but not so cautious, about interpreting for the public what should be regarded as acceptable.

'The criterion we (in the Public Health Service) have been using is that if there is some 10-20 percent fluorosis in the community, that would not be objectionable, because in those places the degree of intensity is not greater than the accepted designation of "mild".' This splendid sentence – a kind of scientific-sounding way of biting your own tail – has become yet another anti-fluoridation classic. Up to 20 percent dental defects in the community is called 'mild' – *ergo* it can't be objectionable.

To this M. C. Smith and H. V. Smith, two well-known early fluoride researchers, said in 1942:

'There is ample evidence that mottled teeth, though

they be somewhat more resistant to the onset of decay, are structurally weak and unfortunately when decay does set in, the result is often disastrous.'

According to Exner, the State Dental Health Officer for California, a Dr Richards, in 1967 stated that 'mild' mottling was 'desirable from both a caries resistant and esthetic point of view' and that teeth that are not disfigured thus should be called 'fluoride deficient' instead of 'normal'. The term 'fluoride deficient' has been taken up with enthusiasm by British fluoride promoters, and indeed by promoters anywhere, though there is no scientific justification for such semantics. It is an appalling and unethical way of misleading the public into thinking that they're missing something they ought to have, and has been capped by Professor Frederick Stare's personal invention: 'mineral nutrient fluoride'. If you accept this kind of logic for what it is, you end up thinking that *un*-mottled teeth are 'deficient' because they lack a 'mineral nutrient'.

Mottling is, in any event as we have seen, no guarantee that you will have no cavities.

Smith and Smith had further comments when they did a survey of St David, Arizona, back in 1940. With a level of 1.6–4.0 ppm, they found that 'caries, once started, evidently spreads rapidly'. Repair was highly unsuccessful, and there were more than 50 percent of false teeth in the age group 24-26. 'Very rarely adults were found whose teeth, though mottled, were free from caries.' This inseparable connection between mottling and fluoride is inescapable, and the margin between the initial benefit and caries protection and eventual mottling and decay is a very narrow one indeed.

In Sweden, at the time when they decided to ban fluoridation, there was much talk of 'black teeth' and press reports that in Billesholm, which has a high natural fluoride content, young people of 16-17 had all their teeth capped because of 'yellow' discoloured teeth (though the dental authorities in the area have been unwilling to comment on these claims).

In 1962 in the *Journal of the American Dental Association*, A. L. Russell reported 50 percent 'opacities' among continuous negro resident children in Grand Rapids, Michigan and 29 percent among those born elsewhere, compared with 36 percent among continuous resident white children and 28 percent among those born elsewhere. 7.1 per cent of white and 15.9 per cent of negro children showed 'positive signs of fluorosis'. Thus those very children most in need of dental 'protection' – the poorest members of society – are the ones who are also most exposed to the dangers or drawbacks of fluoridation.

And if these groups – as it is thought – do not buy or use enough toothpaste, that too may have its saving grace since Ralph Nader has claimed that fluoride toothpaste may stain rather than clean the teeth. (In any case fluoride in toothpaste is known to work best if the paste is a very abrasive one – and the abrasiveness of toothpaste is another dental feud in which passions run high.)

The late Adelle Davis, a biochemist with the University of Southern California's Medical School and a well-known nutritionist on the health food side of the fence in the US, stated:

'In 1962 our Public Health Department reported that the children in Newburgh, New York, . . . had slightly more decay after fluoridation than before. In Baltimore, Maryland, where water has been fluoridated since 1952, rampant decay has steadily increased. In Puerto Rico not only has tooth decay increased, but 64 percent of adolescent boys have mottled and permanently stained teeth from fluoride excess.'

This Puerto Rican situation was of course what got Dr Monteleone into hot water, when he went down to have a look at the 'beautiful teeth of Puerto Rican children whose water supply has been fluoridated for 12 years'. Instead he was 'startled' to observe rampant dental caries, mottling, gingivitis, and malocclusions. But he noticed that the *older* (and pre-fluoridation) Puerto Ricans all seemed to have strong, healthy teeth.

But it is the Newburgh-Kingston study which keeps

cropping up, as Ziskin warned at the beginning. In Norway this study in particular has been criticized by Professor Per Ottestad, of the Norwegian Agricultural College.

'No one who has been involved with sound scientific research and who has taken scientific methods seriously could possibly accept the so-called scientific basis for the promotion of fluoridation.'

Ottestad, like many others before him, finds that there still are really only four studies on which a recommendation to fluoridate could be based: the classical North American ones of Newburgh-Kingston, Grand Rapids and Evanston, and Brantford-Sarnia in Canada, even though the control cities for Grand Rapids and Evanston disappeared from the picture when they were fluoridated in the middle of the trial.

In his analysis of Newburgh-Kingston, he cites a letter from John S. Forst, a New York health official, to Dr James G. Kerwin, New Jersey – just one of the many private letters with information which disputes the published claims for fluoridation and which are rarely published or brought to the attention of the public. In this case the letter revealed the fact that the percentage of schoolchildren with dental defects was higher in fluoridated Newburgh (63.4 percent) than in Kingston (41.6 percent). The defects included caries, periodontal disease and malocclusion. Information which, one would have thought, the public had a right to know.

In Austria, detailed statistical analysis has come from Dr Rudolf Ziegelbecker of the Institute for Environmental Research, Graz. Ziegelbecker has also pointed out that figures published for fluoridation in the German town of Kassel 'proved' the efficacy and safety of fluoride in children 'who did not get the fluoridated drinking water because the aqueduct led into another part of the town without the investigators knowing it'.

Ziegelbecker, despite the high standard of his statistical analyses – is one of the scientists whose papers have been refused publication by, for instance, the *Journal of the American Dental Association.* This journal has also turned down letters from Professor Schatz, not with serious scientific objections but because he should not be allowed to use the *Journal* as a 'platform for his anti-fluoridation views'. One may ask, if scientific journals are not to be a platform for differing views, where, indeed, are scientists with differing views to be provided with a platform?

One might also ask, how trustworthy is a journal that feels its position to be so shaky that it cannot risk publishing even letters from dissenters?

It is often stated that some 10,000 (or 3,000 or 28,000) scientific papers have been published on the benefit and safety of fluoridation. It would be more correct to use the word 'publication' than 'paper' – a lot certainly are not 'scientific papers'. The actual number varies – like the percentages – not a little according to the eagerness of the speaker. Certainly a lot has been published on the subject, but there are relatively few good controlled studies, and much of the literature on fluoride is on toxicity and various medical aspects and could not be said to support the claims for fluoridation.

The word 'controlled' is suspect too, since fluoride promoters do not use it in its strict scientific sense, but in a general way. When they talk about 'controlled fluoridation' they mean that the amount of fluoride added to the water is controlled, i.e. not too much and not too little. Practically nothing else is controlled.

How little 'control' there was even in the classic studies has been described in detail by Sutton. His book is one long, sorry tale of misquotes, uncertainties about water levels, of not making sure the composition of the water in the control town with regard to factors other than fluoride was similar, of ignoring the excellent standard of dental care in the study town as compared with the unfortunate control, of increase in caries-free teeth in the control town, of suddenly fluoridating the control town in the middle of the study, of alteration in sample size, of forgetting to take population movement into account, of 'weighting' of results where different age groups are lumped together so that differences are hidden, or exaggerated, as may have been deemed desirable – even of not being able to count.

Thus in Evanston the number of children was added up wrongly and given at different times as 4,375 and 3,692 and 3,310. After Dr Sutton had pointed this out, it was admitted that only the last one was correct, but the fluoride promoters did not seem to think that a thousand children more or less should give cause for statistical worry. (Even in the British 1969 HMSO report there are examples of the authors getting their own simple percentage calculations wrong).

Sutton lists his objections:

Odd experimental and statistical methods;

Failure to consider random variations and examiner variability, and to eliminate examiner bias;

Omission of relevant data;

Arithmetical errors;

Misleading comments;

Controls were either doubtful or inadequate;

No control was employed in one trial;

The published data do not justify the statement that caries rates remained the same in control cities;

The sound basis on which the efficacy of a public health measure must be assessed is not provided by these five (there was an extra town involved in the Canadian study) crucial trials.

Some of the points have been taken up by Professor John Polya of the Department of Chemistry, Tasmania University. Polya has had fun analysing the figures put out by the Tasmania Department of Health Services in 1961-62.

The Department claimed a 54-65 percent reduction in caries in fluoridated areas. They quoted in evidence fluoridated Brantford in Canada where 20.68 percent were caries-free (in the 12-14 age group), as compared with fluoride-free Sarnia where only 3.30 percent were caries-free. This would of course also mean that in Brantford 79.32 percent of the children had diseased teeth, and in Sarnia 96.70 percent. 'Therefore the passage (in the text) claims that 1 ppm fluoride reduces the incidence of dental decay from 96.70 percent to 79.32 percent, which is a reduction of 17.38 percent. This is quite impressive, but it's not the same as 54 or 65 percent.'

The Department also quoted Beaconsfield in Tasmania which was fluoridated in 1953. Before fluoridation only 16.6 percent had decay-free permanent teeth, while in 1958 the figure had risen to 52.7 percent.

'It is of interest to note that the decay level of the permanent teeth at this age before fluoridation was between 36 percent and 90 percent greater than the figures for similar age groups in North America before fluoridation,' the Department points out.

Polya works this out:

36 percent above 96.70 percent means either 132.7 percent (adding up the figures) or 131.5 percent (adding to 96.7 percent 36 percent of itself).

90 percent would give either 186.7 percent or 184.7 percent.

'How,' he asks, 'did the children of Beaconsfield get *132-187 percent* of their teeth in a mess? How does one get more than 100 percent of one's teeth good, bad, or indifferent?'

This was only taken from a pamphlet explaining, 'Why your State Health Department recommends Fluoridation', but it does illustrate the remarkable absurdity of the percentage figures which are bandied about. Needless to say, people who actually bother to check such figures are not well liked, and Professor Polya was described to me by a spokesman at Australia House in London as a 'raving maniac'. He has also been described to me by someone who has actually met him, as a quiet gentle man. At least he is better at sums than his Health Department.

Whatever one makes of the fluoridation statistics, all the

other factors which are if anything more important than fluoride, have on the whole been ignored, except in very rare cases.

Professor G. Neil Jenkins – remarkably – in the 1960 edition of his book states about Colorado Springs:

'. . . a prosperous community with a high standard of dental treatment and a water supply containing 2.5 ppm, a low caries rate was apparent even among the 40-50 age group. It is not known whether the greater duration of the effect found . . . is caused by the higher levels of fluoride, by the difference in dental treatment, or by other factors.'

Jenkins himself has done much excellent writing on these other factors. Much, in fact, is known about them, though there has been little attempt to inform the public.

In 1972 Dr Abraham Nizel of the University of Boston, for instance, pointed out that 'personalized nutritional counselling could reduce caries by 50 percent.' Albert W. Burghstahler, Professor of Chemistry at Kansas University, mentions that two surveys were done in 1965 in which breast-fed children living in an unfluoridated area were found to have about the same reduction in dental caries as children living in a fluoridated community who had not been breast-fed. Breast-feeding is certainly known to be important. In babies who suffer from hypocalcaemia as a result of being bottle-fed, the enamel practically crumbles away from the underlying dentine.

It has been shown too that strontium and vanadium present in the water may reduce decay dramatically. Yet no one advocates that breast-feeding or strontium or vanadium should be made compulsory. It is known too that regular chewing of tough food is not only important but the very act for which teeth were made. Jenkins himself comments on 'the underdeveloped condition in which the masticatory apparatus exists in civilized man'. There is a considerable literature on these many other factors.

Yet Professor Frank Lawton, Director of Dental Education at Liverpool University and the then President of the British Dental Association, proudly claimed in June 1973 that 'the only fully proved method of strengthening tooth resistance to caries attack is the incorporation of fluoride into the surface layers of enamel, which can be most effectively achieved by the fluoridation of water supplies'.

And Professor D. Jackson of Leeds University, claims that, 'all scientific truth is for fluoridation . . . nature has proved that natural fluoride in water has a beneficial effect, therefore added fluoride would seem to be essential'.

It isn't only the public who are confused. Some dentists are too. In September 1963, one dentist writing in the *British Dental Journal* accused the British Dental Association and the Department of Health of 'falsification of the evidence' and 'extravagant claims'. And a dental practitioner in Kilmarnock wrote, in a local newspaper in 1962:

'It seems to me that the scheme for fluoridation of water supplies and the optimistically glowing reports thereon, are a convenient red herring for the Government of the day to induce public complacency and detract attention from the fact that they have failed to expand the facilities for training dentists, are neglecting the dental health education of the public and are attempting to obtain a dental health service on a pinch-penny and shoe-string basis.'

The moving aspect of the controversy is perhaps the frequency with which the promoters call 'nature' as their witness, yet decry those other nature worshippers, the health food faddists who simply want you to eat healthy food instead of Mars bars, and who think it is wrong for schools, and waiting rooms in dental surgeries, to have slot machines dispensing sweets.

In any case, even nature gets the mixture wrong.

The *Forensic Science Society Journal* in October 1972 quoted a passage from Treatise on the Law of Evidence, one of the classic legal textbooks on evidence, which contains criticism of the testimony of experts which – it seems – would make an admirable post mortem on the present generation of fluoridation promoters:

'They do not indeed wilfully misrepresent what they think but their judgements become so warped by regarding the subject in one point of view, that even when conscientiously disposed, they are incapable of forming an independent opinion. Being zealous partisans their Belief becomes synonymous with Faith as defined by the Apostle, and it too often is but the substance of things hoped for, the evidence of things not seen.'

IV

Safe as Mother's Milk

'Allegations are made from time to time that people have been made ill by consuming fluoridated water.' This pathetic statement in the 1969 HMSO report is matched by Professor Yngve Ericsson's remark in his introduction to the 1970 WHO monograph:

'Of the fears of harmful side-effects from controlled fluoride ingestion that have often been voiced, some can be regarded with the utmost scepticism on account of their unrealistic nature.'

Jenkins wrote in the March 1973 issue of *Medical Digest*:

'The few gaps in knowledge include the amount of maximum intake for heavy tea-drinkers in Britain, and the effect of a life-long intake on the vague symptoms, such as aches and pains in joints, in old age.'

So this is what it amounts to: aches and pains in joints in old age. A few doddering seniles against the distinguished gentlemen who have proved that fluoride is absolutely safe, is a mineral nutrient, is essential to life, and such touching remarks as the one which came from Samuel L. Andelman, Commissioner of Health in Chicago, 'The fact remains that fluoride in a concentration of 1ppm is as harmless as mother's milk'.

In a country where motherhood ranks with the Flag and the wisdom of the Founding Fathers, what better proof (even if they don't breast-feed)?

And yet Professor Ericsson – according to Bergström – claimed at a meeting in Stockholm in January 1969 that infants fed powdered-milk foods dissolved in water fluoridated at 1ppm, would ingest approximately 30 times more fluoride than breast-fed ones.

One may argue about whether this should be a cause

for concern or for rejoicing. What is clear is that safety is a relative concept.

It cannot be easy for the lay public to decide whether fluoride is a 'poison' or a 'mineral nutrient'. T. W. Davenport, a Tasmanian JP, has complained that in Tasmania at the time of fluoridation, fluorides were removed from the Poisons List. The Tasmanian Pharmacy Board said it was done on the authority of the WHO. The WHO later denied all knowledge of it. In Britain, fluoride additives are allowed in baking powder and self-raising flour and in small amounts in ingredients used in the preparation of food 'consisting wholly or partly of an acidic phosphate' – and otherwise not. As far as the law is concerned *drugs and water* are not considered to be 'food', and are exempt from these restrictions. Drugs are, of course, otherwise restricted and controlled. Water intake is not. The fluoride pollution of food also is not.

'The deliberate addition of those poisonous substances, sodium fluoride into public water supplies, and of stannous fluoride in popular toothpastes, with the intent of delaying the onset of dental decay is a most unscientific and unethical measure,' says Dr Reginald Holman, of the Welsh National School of Medicine's Department of Bacteriology.

We should distinguish here between different types of fluoride.

On the level of basic chemistry the fluoride ion is obviously the same wherever it comes from, and it has been argued that consequently one should not distinguish between 'natural' (calcium) and 'artificial' (sodium and others) fluoride. However, it is known that physiologically the body copes better with the effects of fluoride if calcium is present. Indeed, calcium fluoride is commonly regarded as safer because of its low solubility. An editorial in *The Lancet* in August 1960 summed up the situation briefly:

'Elemental fluorine is a highly reactive substance and does not occur free in Nature, but the fluorides, both simple and complex, are found almost universally. Many of them are only sparingly soluble in water, and their toxicity is related to their solubility and their degree of dissociation – hence, sodium fluoride is more toxic than the corresponding calcium salt.'

As already pointed out, the fluoride encountered in 'natural' drinking water is calcium fluoride, and the water is hard, while soft water tends to have very little fluoride. 'Artificial fluoridation' is therefore a different situation in that sodium fluoride, sodium silicofluoride, or hydrofluosilic acid is added to usually soft city water. Any calls on nature to prove the safety of the latter by comparing it with the former just won't hold.

What happens once the fluoride is inside the body is more than complicated and the arguments about it would fill volumes. It is, as Ericsson puts it, 'the most exclusive bone-seeking element existing, owing to its great affinity to calcium phosphate'. Put crudely, one might illustrate the problem with an example given in the *Canadian Medical Association Journal* in 1959 and again in 1960 by Dr William A. Costain – admittedly an antifluoridation man, but one of those the public doesn't usually hear about.

Costain reported that he had found that everyone of his patients given three mg of sodium fluoride a day suffered ill effects, while no one given 25 mg of calcium fluoride suffered ill effects. The symptoms in this case varied from bladder irritation to mental disturbances, deterioration of the skin and falling out of the hair. I shall return to the many different complaints justly or unjustly linked with fluoride presently. What matters in the first instance is the fact that nature is the wrong witness when discussing what the antis call 'artificial' and the pros 'controlled' fluoridation, and that the authorities have largely turned a blind eye to this aspect of the controversy, even those authorities which are there to safeguard us in exactly this sort of situation.

Professor Arvid Carlsson, head of the Department of Pharmacology at Gothenburg University – and a leading Scandinavian antifluoridation expert – has pointed out.

'It is not worthwhile to conceal the fact that it is a question of applying a pharmacologically active substance to an entire population in order to induce a systemic effect.'

Another Swede, toxicologist Stig Ingvar Karlsson, wrote in the *Swedish Medical Journal* (Laekartidningen) in 1969 that the fluorides should be treated as a greater risk than cyanide.

Professor Douw G. Steyn, Emeritus Professor of Pharmacology, University of Pretoria, told a meeting of the South African Medical and Dental Associations in 1969 that it had always puzzled him that the FDA rule that the maximum allowable concentration of potentially toxic chemicals in food and beverages should be one hundredth of the smallest quantity of the chemical which induced disease in 50 percent of experimental animals, i.e. with a safety factor of 100, for most chemicals – but with a safety factor of only 3 in the case of fluoride.

With official attitudes to the safety of fluoride as lax as this, the safety can perhaps only be discussed in the light of specific diseases connected with the substance. These, far from being only a question of aches and pains in old ladies, include cancer, goitre, mongolism, kidney disease, and osteoporosis, with additional problems caused by organic fluorides used in anaesthetics gases and aerosol propellants and even glue and connected with heart failure.

All of them are controversial, and all of them are difficult to assess.

The reason they are so difficult to assess, is simple. John Marier in Canada puts this very clearly in his report, *Environmental Fluoride*: since no Canadian hospital or medical laboratory routinely conducts diagnostic fluoride analyses, and considering the variability of the effects of chronic low-dose fluoride ingestion, 'it is difficult to document the effects of fluoride on man in Canada'.

That quote could be applied to just about every country in the West that has ever advocated fluoridation. Put a little more crudely one might say that you don't find what you don't look for – and this holds true for much of modern medicine except gross and obvious illness and epidemics.

'It took 75 years to suspect that phenacetin caused renal disease, 30 years that amidopyrine caused agranulocytosis and 10 years that tetracyclines disturbed bone development. This should be proof enough that the medical profession does not quickly discover what it does not look for,' says Professor Carlsson. He goes on to quote H. C. Hodge and F. A. Smith – two wellknown American fluoride researchers – as saying:

'No specific large-scale epidemiological studies are available comparing the health of fluoridated communities with that of communities where the water supplies contain only traces of fluoride.'

Marier, commenting on kidney patients, quotes, in connection with one American epidemiological survey in 1958:

'Those with chronic illness and diseases known to affect bone structure were excluded. Thus, there were none with primary or metastatic bone cancer, longstanding cancer of other organs, or renal or parathyroid disease.'

Even with the best will in the world, it would be quite impossible to concede that the fluoride promoters have been overkeen to find out whether fluoride was really harmful.

Burgstahler says:

'Unfortunately, owing to widespread assurances, especially by public health officials, that there are no demonstrable ill effects from fluoridation, comparatively few physicians are even aware of the symptoms of incipient nondental toxic effects of fluoride. Illness from fluoride in drinking water usually develops over a number of years and is thus easily mistaken as being due to other causes which, to complicate matters, may also play a role in such illness.'

Professor Steyn has also repeatedly pointed out that

ordinary doctors and dentists simply haven't been properly informed about which symptoms to look for, and adds:

'From the chronic toxic point of view the F atom could be considered to be the most dangerous poison when taken in small quantities un-interruptedly over very long periods, as it has a broad-spectrum action on enzymes which play a very important role in normal physiological processes in the body. Symptoms could appear as long as 20-30 years, or longer, after ingestion of fluorides, as fluoride is so very prone to accumulate not only in the bones and teeth, but also in the soft tissues of the body.'

When expert toxicologists hold such opinions, one should not perhaps be surprised that they insist so obstinately on opposing fluoridation, as Professor Steyn has done.

Even the WHO report admits, that 'we are still a long way from understanding precisely the biochemical mechanisms underlying the role of fluoride in biological systems' and 'little is known about the *in vivo* effects of fluoride on enzymes and the various facets of general metabolism of the living organism . . .'

The report even states,

'Since only trace amounts of the element are required by the organism . . . animals and human beings are probably rarely in acute need of it. It is also possible that our present state of knowledge concerning optimal levels and essential functions may be quite inadequate.'

And that from the report which is most often quoted in favour of fluoridation!

B. C. P. Jansen, a pro-fluoride Dutchman, and Emeritus Professor of Physiological Chemistry at Amsterdam University and a former head of the Netherlands Institute of Nutrition, says:

'In order to reduce the amount of dental caries man must be exposed to possible fluorine intoxication.' He then asks how long man can imbibe the extremely toxic fluorine without having damaging results and what dosage can be tolerated one's whole life? He seems to think that nobody knows.

With such statements coming from the pros, the antis really don't have much to add, though of course they will continue to do so.

Professor Sir Arthur Amies, Emeritus Professor of Dental Medicine and Surgery, University of Melbourne, and a co-author with Sutton, has stated:

'Fluoride, even in very minute concentrations is strongly toxic and acts as a protoplasmic poison. . . . the real issue is not what is the immediate toxic dose, but what effects are to be feared when a small dose is taken over a period of years . . .'

He adds:

'The United Kingdom Mission which studied fluoridation in the USA (1953) recommended that in centres used for trials of fluoridation the proposed studies " should include full medical and dental examinations at all ages". I know of no full continuous medical studies which have been carried out anywhere.'

Geoffrey Dobbs, a senior lecturer in forest botany at the University of North Wales, and an anti-fluoridation man, has criticized the way in which the British health authorities originally adopted a policy of fluoridation after the 1953 Mission and further after correspondence in the *British Medical Journal* in 1955.

The decision to further justify the *safety* of fluoridation was based on an American survey. In this survey Barlett, Texas, had been selected because of a high-fluoride water content of 8 ppm, with a control town, Cameron, with 0.4 ppm. 116 and 121 subjects in each respectively were examined. During the ten years of the study 14 people in high-fluoride Barlett died, as against only four in Cameron. This was regarded as too small a sample to have statistical significance. Consequently – and somewhat oddly – it became accepted that the high fluoride level was safe because the figures showed no statistical significance – though the mortality rate in the high-fluoride town was three and a half times that in the low-fluoride town. In 1957, according to Dobbs, the American Medical Association re-examined the figures and decided that they were just within statistical significance.

Either way the sample is very small and does not justify sweeping claims. The British authorities, however, accepted this evidence as 'the strongest possible evidence' of the safety of fluoridation, (except, they state in the *British Medical Journal,* for dental mottling in the 'fluoridated' area, though, again, the high fluoride level in Barlett was natural).

The Mission in its report in 1953 even repeats the famous mistake which occurred in the original printing of the story of the death of a 22-year-old Texas soldier, attributed to fluoride, who had lived for seven years in a region of 1.2 ppm. This was mistakenly printed as 12 ppm – and has been mistakenly quoted many times since. Only *The Lancet* seems to have got it right, and indeed *The Lancet* deserves some credit for not being quite so easily taken in as some other sections of the medical press.

Lastly, before dealing with individual diseases, one of the favourite 'proofs' of the safety of fluoridation is the amusing example of the BATH, a story which originated in the US and has been quoted over and over again for years. The challenge was taken up by Dr Hugh Sinclair, a nutritionist formerly of Magdalen College, Oxford, in a marvellously sarcastic letter to the *British Medical Journal* in February 1968.

There appears, he says, to be confusion between the Ministry of Health and Mr. Stewart Ross, Chairman of the Dental Health Committee of the British Dental Association. The Ministry, in answer to the question, 'Is fluoride a poison?' had answered, 'not at the level used in fluoridation. . . . It would be necessary to drink at one time two and a half bathfuls of water containing fluoride at a concentration of 1ppm before any harmful effects due to fluoride would be experienced'.

Moreover, Mr. Ross himself had said at a British Dental Association conference in Oxford, that it would be necessary to drink two and a half bathfuls of water *per day* to ingest a toxic amount.

As usual it was left to the antis to check the actual figures. This Dr Sinclair did.

'I interpreted this as being *acute* toxicity,' he commented, 'and the Ministry has kindly confirmed this, although the relevance of acute toxicity is not obvious to me unless it be assumed that a person will drink deep of the Ministry's doctored water only once.' (For an unmedical reader, *acute* means an immediate condition, i.e. the dose which will kill you there and then, while chronic means the long-term effect.)

Inquiries and measurements then revealed to Dr Sinclair that to fill an average bath with water would take 10 cubic feet or 283 litres, and therefore two and a half bathfuls of water at 1ppm fluoride would contain one and a half *grammes* of sodium fluoride. Forty-five percent of this would, of course, be fluoride. And yet the British Dental Association had stated that only 125 mg of fluoride (about 276 mg of sodium fluoride) would produce acute non-fatal symptoms and 20 mg fluoride a day (about 44 mg sodium fluoride) would produce chronic poisoning.

'In so momentous an issue as dosing daily millions of persons with fluoride, surely the spokesman for the British Dental Association would not make the elementary error of confusing acute and chronic toxicities?' Dr Sinclair asks, and also suggests that Mr Ross 'put to the test the truth of his statement by ingesting 1¼ grammes of sodium fluoride daily with his mid-day meal for six months?'

We do not know Mr Ross's answer, but it is unlikely that he would have taken up the suggestion, even though the immediately fatal dose of fluoride is supposed to be 2 grammes and over. (Some give it as 5 grammes, but more usually it is thought to be between 2 and 10. For obvious reasons one cannot put it to the test on humans. The WHO report gives advice about what to do should a child ingest large quantities of fluoride tablets or fluoride toothpaste. A 50 g tube of toothpaste may contain as much as 200 mg fluoride. However, I have not come across reports that accidents have happened in this way. Anyway, parents with small children should not regard fluoride as a safe 'mineral nutrient' which their children can gobble up as they please.)

In 1967 The International Society for Research on Nutrition and Vital Substances – which has a membership drawn from 76 countries, 65 percent of them medical scientists, and quite a few Nobel Prize winners – passed a resolution which read:

'The Scientific Council of the International Society for Research on Nutrition and Vital Substances recommends all governments, state parliaments and city councils, who concern themselves with the problems of the fluoridation of drinking water and the protection against dental caries, to refrain from the fluoridation of drinking water, which is in reality a medication, as long as the scientific aspects of this problem will not be satisfactorily clarified.

'The fluoridation of drinking water releases a fluorine circuit which includes vegetables, fruit and other horticultural products as well as milk, and has an uncontrollable effect on the human organism.

'The fluoridation of drinking water should not be valued according to briefly occurring successes, which are judged very differently, but rather according to the later hazards which are then incurable.'

The year before, the FDA had hastily and suddenly banned the use of fluoride in prenatal drugs, though they gave no advice on how pregnant women might avoid it in drinking water. As we shall see in the next chapter, there are good reasons for exercising caution.

Since a standard technique of the fluoride promoters is to counter reasoned statements by distinguished pros with unreasonable ones from the antis, it might be permissible once in a while to do the opposite. To all these scientific warnings, what do the pros have to answer? The famous leaflet from the Pennsylvania Department of Health gives detailed advice about what to do and say.

'Rather than debate the points the opposition raises,

take the offensive. Never let the antis state something against fluoridation, and then catch yourself answering them. Keep them on the defensive by:

'1. Expose the nature and tactics of the opposition. Show their backgrounds. Who they are. The purpose of their fight. How they stand to profit.

'2. Ridicule them. The best remedy for the fear technique, is ridicule. Try to get people to laugh at the antis. . . . Example: The antis shout "poison" – Counter with "the people in Camp Hill are falling over like flies because they have fluoridated water".'

The remarks even by British fluoride promoters would echo such attitudes. As a measure of the standards set itself by a group of medical scientists and medical practitioners, this is the sort of thing the unsuspecting public needs protection against – even if fluoride really were absolutely safe and even if only as a protection against being treated like idiots.

On the whole it could be said that the stand on safety and toxicity is based on the argument that since fluoride is natural, it should not be necessary to apply to it the same stringent rules used with other food additives.

The 1969-70 circular to British Medical Officers of Health states that 'If an intentional food additive is a constituent of the diet or is a normal body constituent this may provide grounds for a lower safety margin' and 'it would not be feasible to apply 100 times the safety margin to many common food additives, e.g. sodium chloride (table salt)'.

It almost sounds as if the Department of Health equates fluoride with ordinary salt. Moreover, the judgement of safety has been based on natural fluoride and natural levels, disregarding the role of diet and especially calcium, and disregarding that naturally occurring calcium fluoride is not the same as adding sodium (or other industrial) fluorides to soft water.

The circular adds:

'At the time the US Public Health Service decided to adopt fluoridation as a policy, evidence of safety was certainly not as complete as it is today. Nevertheless there was a considerable knowledge derived from observations from areas where fluorides occurred naturally in water supplies and it was felt in the US that this provided evidence to proceed with fluoridation and at the same time to *conduct further research to confirm the safety of the procedure.*' (my italics).

But what if the safety was not confirmed? Having endorsed fluoridation and being determined to continue it, it seems that there may have been at least a political necessity to make sure that the safety was confirmed and not challenged.

It is in the light of this attitude that the various diseases which have at one point or another been connected with fluoride, should be seen.

V

'Aches and Pains'

Bone diseases (osteoporosis)

In 1962, a patient suffering from osteoporosis and treated with sodium fluoride at King's College Hospital, London, staggered into the out-patients' department, blind in one eye and with impaired sight in the other as a result of optic neuritis caused by the treatment.

This one incident was reported in the *British Medical Journal* in August 1963 – and to this day mention of it keeps cropping up in anti-fluoridation literature from all over the world.

It was, of course, an isolated case, and the paper was a brief one. But it was not short enough to prevent confusion, because it was accompanied by an editorial talking in terms of doses of 60 mg a day, while the patient had only been given 20 mg a day and the authors state this quite clearly. The writer of the editorial seemed to be keen to reassure the readers that this admittedly drastic case had no direct connection with the policy of fluoridation and the small amount added to drinking water.

The authors of the paper – M. G. Geall and L. J. Beilin – were cautious enough to state that although it could not be definitely established that the eye lesions were caused by sodium fluoride, there was no evidence for any other cause.

'. . . the fact that his eye trouble began only six weeks after starting to take a powerful enzyme poison makes the later highly suspect.'

They also point out, however, that as his parathyroid glands had been removed, it might be possible that less fluoride was laid down in bone and more in soft tissue and this might be the cause of the unfortunate state he found himself in.

But such details are not for laymen. A man goes blind because his doctor gives him rat poison – it is the sort of case the antis will jump at and it seemed that they had good reason. In the correspondence which followed in the *British Medical Journal*, Dr George Waldbott, a well-known American antifluoridator and dermatologist and editor of the journal, *Fluoride*, pointed out that the originators of the fluoride treatment for osteoporosis had admitted as early as 1944 that this treatment was 'playing with fire'.

Today, Professor John Anderson, head of the Department of Medicine at King's, says that he still has his doubts about the effectiveness of fluoride in osteoporosis and that he regards the recommendations for this treatment as 'unscientific'. A polite medical way of saying that he doesn't like the idea of it at all. And yet Professor G. Neil Jenkins has publicly stated, as late as 1972, that fluoride is good for osteoporosis and fluoride promoters claim that there is less of this disease in high-fluoride areas. The same sort of claims were repeated in the WHO 1970 report.

The claim goes back to, of all people, the famous or infamous Frederick Stare at Harvard. Frederick Stare has done one of the most oft-quoted and most widely publicized medical surveys in recent history. News of it was released even before it appeared and it was reported and written up in the popular media across the States. The *Detroit News* put it on its front page, perhaps because a referendum on fluoridation was just about to be held in that city at the time. It is still being quoted by medical researchers in good faith even today.

In 1966 Frederick Stare, together with Dr Daniel Bernstein who had been trying to treat osteoporosis with large doses of fluoride, went out and examined the incidence of the disease in North Dakota and compared this with the fluoride levels in the drinking water. The paper containing the results was published in the *Journal of the American Medical Association* on October 31, and Stare and Bernstein claimed that they had proved that fluoride was good for osteoporosis because there was less osteoporosis in high-fluoride areas. Unfortunately for Stare, the antis claim that some 'high' areas were listed as 'low' and some 'low' ones as 'high'.

The antis have never let him forget it, and he has never, as far as I know, commented on their accusations nor has he denied them. On the contrary, it seems that he continued to repeat the mistake.

It may of course have been a genuine mistake. But the antis are going to be hard to convince. The year before the North Dakota paper appeared, Stare broadcast on the radio and when asked:

'Has there ever been a case of fluoride making a person ill?' he answered:

'Not a single case.'

It seems that the *Journal of the American Medical Association* subsequently refused to publish criticisms of the North Dakota survey. Several appeared elsewhere, but the dispute has by no means been settled yet. Partly, or perhaps exactly because, these criticisms did not appear in the journal that carried the original paper, newcomers to the field often do not know of them.

Basically, the details of the accusations as outlined by F. B. Exner, a leading American antifluoridator who has analyzed the figures, are as follows:

The methods used in the X-ray evaluation of bone density in the patients were unreliable for various medical and technical reasons. There were statistical errors. Some of the subjects did not use the water which was cited, i.e. the water quoted were municipal water supplies but many of the participants – farmers for instance – used their own wells. In fact, in 1937 prolonged drought had lowered ground water levels to such an extent that it had become necessary to drill deep new wells, some of which carried sufficient fluoride to cause mottled enamel. This was in the 'low-fluoride' counties cited and some of the wells had up to 3 or 4 ppm, which is hardly 'low'.

Stare had stated the 'high-fluoride' areas to be in the

south-western part of the state where the level ranged from 4 to 5.8 ppm. In fact it is claimed that the real 'high' area was in the south-eastern part where as much as 9 ppm had occurred.

Two of the 'high-fluoride' areas, Mott and Hettinger, were given as 5.5 to 5.8 ppm respectively – but in the official US Public Health Service register were given as having only 1.2 and 2.2 ppm. The 'low' area of Grafton, given as 0.15 to 0.3 ppm, was officially listed as 0.9 ppm. In 1938 another official report gave the water from wells in 'low-fluoride' Grafton as between 3.5 and 4 ppm. In 1959 another well was added and the level of the pooled waters reduced to 2.8 ppm. In 1964 the source of water had been changed to the Park River, with 0.4 ppm, and during the transition period it was found to be 0.9 ppm. All this is, of course, important when you wish to measure the effect of fluoride in permanent residents over a lifetime.

One 'high' area, the town of Reeder, originally had 0.2 ppm, but later had a new well with 3.5 ppm, and another 'high' town, Regent, was listed as 0.7 ppm in 1959, but 1.1 and 1.5 in 1964. The town of New England had 0.8 ppm in 1964. Apart from that, according to Dr F. B. Exner, there were no wells in the 'high' areas that contained as much as 0.7 ppm.

Another researcher, Philip E. Zanfagna, pointed out in a letter to the *Medical Tribune*, New York, a few months later:

'. . . in fact in the high-fluoride area that was contrasted with a similar low-fluoride area, there were only 80 people of both sexes over the age of 65 studied, a figure too low to have any statistical value.'

No attempts, he said, were made to determine the total daily dose of fluoride. Moreover, natural fluorides are found in hard waters which contain greater amounts of calcium, magnesium and phosphorus, and consequently information from surveys done in hard-water areas have little significance with regard to what might happen if you add a different type of fluoride to soft water.

A Dr Lewis B. Barnett, he said, an orthopaedic surgeon from Deaf Smith County, Texas, had reported in a talk before the Texas Medical Association that he found less osteoporosis in individuals drinking naturally fluoridated water, not artificially fluoridated water. He had made careful analyses of soil and water and concluded that the only mineral from such waters truly important in building resistance to osteoporosis is magnesium phosphate.

Dr Albert P. Iskrant, chief of the US Public Health Service's Epidemiology and Surveillance Branch, Division of Accident Prevention, had also found no evidence of reduced incidence of osteoporosis among individuals drinking artificially fluoridated water, though again there was a reduction in natural areas.

One writer, R. Fulton, of Toronto, wrote at this time about the editor of the *Journal of the American Medical Association*:

'. . . Dr J. H. Talbott has been a promoter of fluoridation ever since he supported fluoridation in the campaign to put it across in Buffalo, NY. He has refused to publish all data unfavourable to fluoridation and has written and published emotionally-toned editorials promoting fluoridation.'

Since then other objections have come in thick and fast. In Britain, Professor B. E. C. Nordin, of the MRC Mineral Metabolism Unit at Leeds University, has said that 'the American study was extremely inaccurate and did not use any objective quantitative method and that in at least one of the areas the concentration of fluorine (4 to 5.8 ppm) was, if not necessarily toxic, certainly above accepted safety levels.'

A survey done in England in West Hartlepool and Leeds led Professor Nordin to the conclusion that there was no significant difference in numbers or degree of osteoporosis in high and low areas.

In Sweden, strong protest has come from Professor Olov Lindahl, a well-known orthopaedic specialist at the Regionsjukhuset in Linköping. He regards the use of sodium fluoride in the treatment of osteoporosis as 'very dangerous' and mentions spontaneous fractures in some of his patients who had been given this treatment.

In June 1970, a paper was written in the *Journal of Nutrition* by no less than ten leading specialists from Cornell University's New York State Veterinary College, and the Graduate School of Nutrition, from Colorado State University's Department of Chemistry, from the Department of Orthopaedic Surgery at the Karolinska Institute in Stockholm, and from the National Research Council of Canada and the University of Ottawa's Faculty of Medicine. John Marier was one of the authors. The study was of fluoride and 'Nutritional Osteoporosis' in dogs, and the conclusion was that calcium deficiency results in the most pronounced osteoporosis, that mineral loss becomes more severe with increased levels of dietary fluoride, and that the proper treatment is correct calcium and phosphorus nutrition – not fluoride addition.

Researchers from the Mayo Clinic Department of Orthopaedics and the Division of Endocrinology wrote in the July 1972 issue of the *American Journal of Medicine* that while fluoride administration has been shown to stimulate new bone formation, the bone is poorly mineralized, and moreover that dietary phosphate, vitamin D, and calcium are more important in treatment.

It is accepted that fluoride does stimulate new bone formation, though the exact effect or its cause is not established. New research seems to indicate that the effect of fluoride in young animals is to cause rickets, while in adults the prevailing skeletal change is the presence of abnormal amounts of bone or simply abnormal bone. Fluoride also seems to reduce the resorption of bone, though not everyone agrees on this point. Dr J. M. Faccini at London's University College Hospital Medical School thinks that while fluoride-containing bone is resistant to even the normal processes of resorption, this is associated with increased resorption of non-fluoride-containing bone, and this is again linked with the effect of fluoride on the parathyroid glands.

There may also be unexpected findings. Thus Researchers at the Tuscon Medical Center in Arizona reported in 1971 that they had found mysterious new giant cells in bone marrow of patients on high fluoride treatment combined with aspirin, though the main effect of this seemed to be gastro-intestinal trouble and bleeding.

An issue of *Fluoride* in October 1970 devoted to osteoporosis showed that newly formed bone in cattle on fluoride supplementation is indeed abnormal and unhealthy bone, and that increased skeletal mineralization following fluoride intake is only the initial phase of a pathological process which is followed by a suppression of calcium absorption from the gastro-intestinal tract. In animals osteoporosis can be easily induced simply by feeding a diet high in phosphorus and low in calcium and to suggest that this disease is caused by lack of fluoride seems totally unacceptable. In Britain, bonemeal tablets have recently been used as a successful treatment at Bristol General Hospital.

Despite all this, the 1970 WHO report states blithely that fluoride should be recommended 'in high amounts during short periods of time for the treatment of osteoporosis', that 'larger doses of fluoride used in the treatment of osteoporosis promote a positive calcium balance and give fair clinical results', and 'fluoride is either eliminated in the urine or innocuously "sequestered" in the skeletal and dental tissues'. And of course the report quotes Stare and Bernstein as proof that living in a high fluoride area is good for osteoporosis. Since osteoporosis is particularly well-documented compared with some of the other diseases connected with fluoride, this is one of the instances where the informed reader cannot help feeling that the authors of the WHO report are deliberately misleading the public. It seems truly appalling that it should be left to *lay* antis to try and draw medical attention to scientific papers which have appeared in perfectly respectable journals and by people working for world-famous universities and institutions – and that these antis should still be laughed at for doing so. In

Britain one would, in the normal course of events, have expected the Department of Health and the British Medical Association to have tried to draw doctors' attention to these famous papers as a warning that perhaps they shouldn't be quite so hot on encouraging their patients to eat more and more fluoride.

Osteoporosis is not the only bone disease which people are arguing about, and to understand how the others fit into the picture, it is necessary to know a little about the diseases themselves. Definition of bone disease in general can be a tricky business, not the least because the experts disagree. Ideally and to be absolutely sure of your diagnosis you'd want to kill the patient and put his bones under the microscope. Failing that, you have to fall back on X-rays and that is often where the trouble starts. Clinical symptoms – aches and pains indeed – are often vague to start with and can easily be confused with totally different illnesses. Briefly those which concern us here are the following:

Osteoporosis, as already mentioned. The word means literally *porous* bone, or to use a more technical term, reduced bone mass or decreased bone density. What this amounts to is that the bone is poorly mineralized and looks like a dry sponge with holes in it under the microscope. The exact causes, say the textbooks, are 'uncertain'. The disease is linked with age, with lack of hormones in older women, with lack of calcium, with 'insufficient formation of the protein matrix' – another typically vague medical hypothesis – with disuse of the bone (all types heal faster with activity than with bed-rest), and finally, with fluoride. Normal bone is continually torn down by the bone-destroying cells (osteoclasts) and built up by the bone-building cells (osteoblasts), and this disease, like the others, is caused by an imbalance between the two processes. Why some people get one type of bone disease and others another, is not at all clear.

At the other end of the spectrum is *osteosclerosis* which

means a hardening of the bone caused by abnormal calcification. This disease has also been called Chalky Bones or Marble Bones because of the patchy appearance under the microscope.

Somewhat different is *osteomalacia*, a kind of 'adult rickets' or 'renal rickets' which may occur in conjunction with kidney disease. The name means *soft* bones, and that is what it is. The bones become flexible and deformed, and the disease is caused by deficiency in calcium, loss of calcium salts in the tissues, and vitamin D deficiency. I shall return to this one later, but a picturesque – if horrible – description of the disease has been provided by Professor D. N. S. Kerr at the Renal Unit at the Royal Victoria Infirmary, Newcastle-upon-Tyne where some kidney patients developed osteomalacia as a result of kidney machine treatment.

One patient had started off 6 feet tall, and then had shrunk to about 5 feet. His chest had caved in, causing difficulty in breathing and finally respiratory failure. He had to be supported if he wanted to sit up or stand because of pain and severe muscle weakness. Such conditions have been linked with the use of fluoridated water in some dialysis units, but in other areas with the same fluoride levels, the trouble, mysteriously, does not occur. (It should be remembered too, of course, that kidney machine treatment is life-saving treatment, and the unhappy choice for the patient in such cases may be between being dead or being alive with osteomalacia).

The most clearly established fluoride-induced bone complaint is crippling skeletal fluorosis, also called chronic fluoride toxicity – a term which agricultural researchers prefer as more accurate. The condition is well-known in cattle and includes osteosclerosis as part of the diagnosis, but in humans early diagnosis can be difficult. The subject does not appear to be taught in medical schools in the West and medical textbooks and dictionaries have only the barest of references to it. Admittedly, it tends to occur as an endemic condition only in the tropics, but it has been known in temperate climates as industrial fluorosis caused by exposure to dangerous industrial fluorides.

There is even such a thing as *'wine fluorosis'*, first diagnosed in Spain, caused by the fluoride preservatives used in some European wines. Researchers at the School of Medicine at Barcelona University recently tried to draw attention to this one, but so far only antifluoridators seem to be aware of it. Wine fluorosis starts as a slow initial osteosclerosis or hardening of the bones, which then gradually resolves into osteoporosis and bone atrophy. As so often when dealing with fluoride-induced diseases, the authors here, too, remark that many cases probably go unnoticed because of faulty diagnosis or unawareness that the disease exists.

In endemic areas with 'normal' fluorosis, there may be dental mottling as an early sign if the fluoride intake has been too high from childhood, but the absence of mottling does not mean an absence of skeletal damage.

In young adults, the only early symptoms are vague pains which are found most frequently in the small joints of the hands and feet, in the knees and in the spine. These may be misdiagnosed as rheumatoid arthritis. In India, a common early symptom is a recurrent tingling sensation in the limbs or all over the body. Pain and stiffness appear next, especially in the lumbar region of the spine. In the end the whole spine may appear like one continuous column of bone. The ligaments and tendons may be calcified, the vertebrae may fuse together. In the end the patient may be bedridden – though mental powers remain unimpaired.

The most detailed research of endemic skeletal fluorosis has been done by the late Dr A. Singh, formerly Principal of the Government Medical College, Patiala, Punjab, India, and S. S. Jolly, Professor of Medicine at the same college.

In the 1970 WHO report, they state that advanced crippling skeletal fluorosis is associated with a daily intake of 20-80 mg of fluoride over a period of 10-20 years, and that this was associated with a level of at least 10 ppm in the drinking water. However, in 1973, at the Fourth Annual Conference of the International Society for Fluoride Research in the Hague, Jolly stated that the incidence of skeletal fluorosis may vary considerably in communities exposed to nearly the same level of fluoride in the water, even at high levels, and conversely many of the early symptoms including limitation of spinal movement may occur where the water is less than *1 ppm.*

The variations seem related to other chemical constituents in the water which are protective against the development of fluorosis, such as magnesium, calcium, and total hardness of the water. Excessive physical stress might also make the condition worse, and men are more effected than women. There are certainly also climatic differences which have not been clearly mapped out. The disease occurs in the Middle East and in Japan, but one could not really draw any direct conclusions as to what might happen in Europe.

Other researchers have studied skeletal fluorosis in India, most recently M. Teotia and S. P. S. Teotia of King George's Medical College, Lucknow, India. These two studied the disease in children and found symptoms such as, 'vague pains, restricted joint movement, backache, stiffness and rigidity of the spine, inability to close the fists, and constipation'. The onset, as always, is insidious. In more advanced cases, the patient's body 'moved as a single unit whenever an attempt was made to straighten it'. They feel – contrary to previous reports – that the disease is common in children and can occur at a very young age, from 11 onwards.

In the West, industrial fluorosis was first described in cryolite workers in 1937 by the famous Danish fluoride specialist, Kaj Roholm, a classic researcher in the field and the only one perhaps who is equally venerated by both pros and antis. In Britain, the Medical Research Council did a survey of industrial fluorosis in aluminium factory workers near Fort William in Scotland in 1949, an enormous work which reached the conclusion that none of the workers examined was found to suffer clinical disability. Traditionally in Europe, industrial fluorosis has been associated with the cryolite worker's 'poker back'.

It is interesting to note that the MRC survey was carried out because of a 'few individuals whose observations in the area had led them to suspect the possibility of a risk to health' (though these were mainly concerned with the local cattle). Since then the health authorities have been less interested in listening to the suggestion of isolated 'individuals', if the suggestions concern fluoride.

There seems to be little writing in the medical journals about industrial fluorosis nowadays and one rather gains the impression that there is no real problem. Part of the trouble may be in the fact that while it was previously assumed that the main danger came from inhalation, the danger today – at least for the general population – is in the ingestion of contaminated food. As the early symptoms so often amount to no more than vague aches and pains, doctors may perhaps be forgiven for not paying any great attention to the possibility of something more serious. On the whole the more serious expressions of fluoride-induced bone disease does not appear to be a cause for worry in the West – but it really is impossible to arrive at any sort of judgement of the situation simply because nobody is looking to find out if the problem really is there.

One of the factors remarked upon in the MRC report – as in a great many others – is the 'remarkable thirst for water associated with fluorosis'. This, the authors think, suggests renal irritation.

Kidney disease

'It should be possible for physicians who have the care of patients with severe chronic kidney disease to recommend the use of imported fluoride-free water should this be necessary', said the Department of Health in 1953 when they recommended fluoridation.

The Department does not appear to have given doctors

further instructions as to when this should be 'necessary', or whose responsibility it should be to make arrangement for importing the fluoride-free water. Surely not the doctors'?

In any event, in the case of universal, country-wide compulsory fluoridation of all water supplies of the sort which the British Dental Association keeps telling the government to introduce – where would the fluoride-free water come from?

The role of fluoride in kidney disease is confused. Fluoride 'in large doses can be classed as a depressant of kidney function', says the WHO report.

But other quotes in the same report only add to the general confusion. The report says that kidney disease does not seem to accentuate the problem of renal handling of fluoride, and that urinary fluoride concentration tends to be lessened in renal failure, but also that there is no trend to indicate that nephritic patients excrete less fluoride than elderly patients with normal kidney function. But then again it states that while fluoridation is undoubtedly safe for persons with normal kidneys, there is a 'remote possibility' that it may aggravate renal disease and that this may lead to a smaller margin of safety than for normal individuals. And yet again it states that the highest levels of bone fluoride found in people with the most severe kidney disease have been within the range for normal people drinking water containing less than 0.5 ppm fluoride.

Different conclusions have been reached by different contributors, so the overall conclusion seems to be: take your pick.

Polya has some sharp comments on American surveys designed to settle this one:

'Even in America the position . . . has not been settled: young fluoridated normals have been favourably compared with unfluoridated inmates of an old age home, and fluoridated children's urine *pooled* in batches of fifteen "did not show abnormalities" ' (referring to a famous test done by McClure).

He described this as 'sharp practice' and asks how you can detect individual abnormalities if you pool the urine?

'Dodges of this kind do not disprove fluoridation but underline the suspicion that all is not safe in the safety racket.'

Strong words. It might be necessary to point out that not *all* fluoride promoters pool their urine samples in batches of fifteen, though the antis never tire of telling each other about the story of McClure, and I have come across another case where the urine was pooled in batches of 12.

The pros firmly believe in the wisdom of McClure and claim that he proved that all that went in, came out. Indeed when you discuss urine and blood levels and kidney function, one has the impression that the argument fluctuates as much as the urine levels. It all depends on whether you wish to prove that fluoride is harmless by being excreted, useful by being retained, or dangerous by being retained and not excreted.

Altogether the body's successful management of fluoride depends on its elimination via the kidneys and the urinary tract. The pros claim that there is a steady state between what is eliminated, and what is retained. This does not mean that what you drink goes straight through the system, but that some is retained in the skeleton while at the same time some previously deposited fluoride is also released from the skeleton, in roughly the same amounts. That at least is the crude theory. Some is also released in sweat and through the lungs, in fact quite a lot may be excreted through the skin. The problem with this theory is that – and this is accepted by both camps – over the years and with increasing age there is an increase in fluoride levels in the skeleton, and also in the teeth and in soft tissue. Consequently even with a reasonable degree of 'steady state', some fluoride still gets retained. It may take twenty years before this small retained amount may become 'pathologically significant' and disease develops, but the fact that some does get re-

retained, however minute the amount may be on a day-to-day basis, is crucial. The pros have often tried to argue that because it can so easily be demonstrated that even large amounts of fluoride are quickly excreted in urine, this must be proof of its safety. I have known one lecturer at a leading London teaching hospital to be quite remarkably naïve about this point. He had, he said, come slowly to the fluoridation thing, but then one day someone got him to take a large dose of fluoride, and lo and behold! It all seemed to go straight through him. As indeed it probably did. He accepted this one little demonstration as convincing evidence of the safety of fluoridation.

In kidney disease the real problem is to try and work out whether the fluoride – if the intake is too high – is the actual *cause* of disease, or whether a too high intake of fluoride is simply a hazard to kidneys which are already diseased. As always in science, the final verdict may be a bit of both.

Reports have come from Canada and the US that fluoridated water is a danger to patients on haemodialysis, or kidney machines, though this has not been confirmed in Britain. Basically, the supposed danger comes from the fact that the fluoride concentration in fluoridated water, at 1 ppm, is six times that in blood serum. Therefore, when such water is used in the 'dialysate bath' of an artificial kidney, fluoride would be expected to move into the patient's blood. As much as 200 litres of water may be used in one session on a machine, representing 200 mg of fluoride. As much as 30 mg of fluoride may be absorbed in the body each time – hence the supposed connection with bone trouble and 'renal rickets'.

However, there are some oddities both in the observations, and in the way they have been reported.

In 1963, a report came from the University of Rochester School of Medicine of a nurse in her early forties suffering from kidney disease who died in a state of convulsive seizures following her fourteenth haemodialysis done with fluoridated Rochester tap water. Two years later a second paper concerning the same case appeared, written by different authors, which surprisingly did not attribute the patient's death to this cause and did not even cite the original paper. The new authors even managed in the text to suggest that their report was the first and only one. D. R. Taves, the main author of the second paper, remarked that there was an excess of bone breakdown and an elevated concentration of fluoride in the bone of the patient, that this could not be proved to be detrimental, and he then – by comparing this kidney case with the 'proved' benefit of fluoride in the treatment of osteoporosis – concluded that the extra fluoride was probably beneficial to the kidney patient. However, the last paragraph still recommended that 'it would seem prudent' to use nonfluoridated water for long-term haemodialysis.

If the antis feel that this way of 're-writing' and updating an original medical report is suspicious, then perhaps they are right. The new report did have important additional information about the ratio of fluoride in the water and in the blood, and the medical significance of this, but the peculiarities about the whole affair still stick out. I myself originally only read the second paper and did not know there had been a previous one. I only found out about the first one because I thought it strange that this apparently harmless case should have been quoted so often in the anti-fluoridation literature.

The challenge was taken up by two Canadians, Dr G. A. Posen from the Ottawa General Hospital, and John Marier, of the Division of Biology at the National Research Council in Canada.

The haemodialysis centre at the Ottawa General Hospital opened in April 1964, and the use of fluoridated water started with the fluoridation of the city's water supply in November 1965. 'Our subsequent therapeutic failure was completely unexpected,' the authors state in the paper which was later presented at the Second Annual Conference of the International Society for Fluoride Research in Barcelona in 1969.

There were ten patients. Two had 'symptomatic renal osteodystrophy' – which means defective bone development connected with the kidney disease – when they were accepted for treatment; six others developed this condition within a year of starting on the fluoridated kidney machines. The disease progressed despite all attempts – including vitamin D treatment – to stop it. In fact, it seemed that vitamin D treatment made things worse.

In October 1968 the hospital deionized all water used for kidney machines, in order to remove the fluoride, and a further report was published by Dr Posen in 1972. Some of the patients in the original survey died, and the antis – as might be expected – concluded that it had been proven that it was the fluoridated water which killed them. That, of course, is an oversimplification; the issue is more complicated than that.

Dr Posen was not as rash, though Marier seems to feel strongly about the danger of fluoride to patients with impaired kidney function. Dr Posen – punished perhaps for the extreme views of the lay antis rather than for his own words – also seems to have been subjected to some abuse by fluoride promoters.

The case still has not been proven, in a general sense.

Professor David Kerr, at the Royal Victoria Infirmary, Newcastle, after his problems with osteomalacia in patients on kidney machines, went into this problem and discovered that in Birmingham where the water is also fluoridated, there are no such problems. It could be that other factors in the water, alone or together with fluoride, cause the problem. Nobody really seems to know. However, there is increasing interest in using deionized water all the same, just to be sure.

Doctors at the Mayo Clinic in Rochester, described two cases in 1972 where impaired kidney function was linked with a high fluoride intake, though – as is often the case with papers on fluoride – they introduced their report by saying that of course the case for fluoridation of water supplies 'is now overwhelming and nearly all commentators seem to agree that the end product is perfectly safe'. As always, it is never quite clear whether this kind of remark is a sop to the pros, an attempt to stall the antis, or a genuine opinion. But doctors reading it must more often than not come away with the impression that everything is all right, regardless.

However, in 1973, even Professor G. Neil Jenkins in Newcastle felt obliged to admit in a fluoridation article in *Medical Digest* that haemodialysis in fluoridated areas 'presents problems'. He reassures us that 'this is being investigated'.

High time too, some might say. The term 'fluorine rickets' had already appeared in the 1949 MRC report on industrial fluorosis. Twenty-four years later, it seems, nobody has got much further.

In Sweden, Professor Arvid Carlsson at Gothenburg University feels strongly about this one and has made some nasty remarks about the attempt in Newburgh in 1965 to analyse 100 children with the intention of elucidating a possible harmful effect on the kidneys from water fluoridation. In this survey 'no specimens were taken if there was any history of clinical illness, no matter how mild, during the previous two weeks'. Carlsson comments that the authors, in fact, excluded from the survey all those cases from which they might have learned something.

In Britain, the only attention given to this problem of fluoride and the kidneys in recent medical 'literature' (at least while I have been looking for it), was a brief report in *World Medicine* on January 31 1973 about the cases at the Mayo Clinic, a misquote about Newcastle in *Medical News* in February 1973, and a note in the now deceased *Science Journal* in March 1967, in which Dr Simon A. Beisler, chief of Urology at the Roosevelt Hospital, New York City, was quoted as saying, 'I just don't feel that fluoridation has been researched the way it should have been. Fluoride in the water can reach every organ in the body and there are indications that it can be harmful over a long period of time'. Dr Beisler does not reply to letters asking questions about fluoride and, on the whole, the medical profession has not shown any great enthusiasm for trying to solve the puzzle.

Heart disease

The question of a connection between fluoride and heart disease is, if anything, even more confusing, and perhaps the weakest case from the point of view of the antis. On the other hand, the claims by the pros that fluoride, or fluoridation, actually prevents heart disease, should be treated with some scepticism.

In general, it is the calcification of the arteries which is of interest, since fluoride has long been known to be part of the 'limestone rock' which deposits on the walls of the blood vessels. It has also been known for a long time that, in the soft tissues, the highest concentration of fluoride is found in the aorta, the main artery.

The problem here – in this long, sorry tale of problems – is: does the fluoride just happen to be there because of its affinity with calcium, or does it speed up the process of calcification for that reason – or does it actually stop calcification by binding with calcium and thus preventing it from being precipitated and turning into limestone rock?

This question brings us back, once more, to Frederick Stare. His North Dakota survey 'proved', he claimed, not only that high fluoride areas produced less osteoporosis, but also less heart disease.

Stare's idea was that fluoride in the water prevents the loss of calcium by the bones and at the same time prevents deposits of calcium in the arteries. In other words, it prevents arteriosclerosis. The editor who recommended the paper, stated in the editorial that 'if these findings are confirmed . . . fluoride will indeed assume an important preventive role in two of the main diseases of ageing – osteoporosis and hardening of the arteries'.

Fluoride promoters have been quoting this ever since and claimed not only that fluoridation prevents heart disease, but also ageing.

On the other hand the antis have quite a batch of evidence, mostly from non-English-speaking countries, (Arabia, India, Japan, and Italy), which indicate that fluoride could 'promote' heart disease. A few reports have appeared in the US. Dr Exner reports the work by researchers at Utah State University where R. A. Call, R. V. Davis, and D. A. Greenwood, between 1957 and 1966, showed that fluoride accumulates in the soft tissue in both people and cattle and that the aorta has by far the highest level, but that isn't really news. In Russia, G. S. Konikova has found that the concentration of free or 'unbound' cholestrol in the blood of persons who had had excessive industrial exposure to fluorides was twice as high as in persons without such exposure.

There is also the curious case of the Philadelphia Zoo. Apparently there was a sudden change in the incidence of cardio-vascular diseases in both mammals and birds in the Philadelphia Zoo coinciding exactly with the fluoridation of the city's water in 1954. Most of the deaths from heart disease were also among the younger animals dying. While the increase in atherosclerosis of the aorta was not striking, the incidences in coronary artery disease, myocardial fibrosis and myocardial infarction all increased several-fold between 1954 and 1958, and went on increasing afterwards.

Attempts to look seriously at the problems, however, have been few and far between, despite early Japanese work republished in the October 1971 issue of *Fluoride.* On the whole, mortality surveys have not been reliable. Reports supposedly showing a higher mortality, including mortality from heart disease, in high fluoride areas, have often been done in areas with a high proportion of retired persons where the death rate would be high in any case. In Newburgh there was a steady increase in the number of deaths from cardiovascular causes from the introduction of fluoridation in 1955 until 1961, but in 1962-63 the trend was reversed.

A survey was done in Antigo, Wisconsin, which was fluoridated in 1949. From that year onwards, the death rate from heart disease increased from an original 359 per 100,000 to 542 per 100,000 in 1970. Throughout this period the US average rate increased only from 340 to 366 per 100,000. However, the survey was done on her own initiative by a Registered Nurse, Isabel Jansen, president of the Citizens' Action Programme – and the scientific and medical establishment are hardly likely to be impressed by such uninvited initiative by an ordinary nurse.

More reliable information – without, I hope, offence to the enterprising nurse – comes from Dr George Waldbott, editor of *Fluoride*. A brief review by him of this and other diseases linked with fluoride appeared in the February 1961 issue of the *Archives of Environmental Health*. However, the question of whether the presence of fluoride in the aorta is 'merely a harmless coincidence' remains. It just seems that not enough people are trying to find out.

There has been one celebrated, and oft-misquoted, case. In 1964, Dr John F. Bacon, of the McFarland Clinic and the Mary Greeley Hospital, Iowa, reported the death of an infant within 16 hours of birth with quite abnormally high fluoride levels in the body, with a huge heart, and with gross calcification in the entire aorta and the region of the heart valves, calcification of the pelvic arteries and of arms and legs extending as far as the wrist and ankle on each side, with marked calcifications in the lungs, and also some calcification in the kidneys and the thyroid.

No explanation was offered. Little was known about the mother except that she was 'allergic to many things', had had recurring urinary tract infection, and suffered from 'mild toxic conditions', whatever these may have been.

Dr Waldbott took up the case and it was further revealed that though the parents had resided for four years in the fluoridated area of Ames in Iowa, there was no evidence of excessive fluoride or water intake. It all remains a mystery.

The significance of the case – in so far as there is any – is in the fact that the placenta is normally supposed to prevent fluoride from getting through to the foetus, though the 'filter' is by no means 100 percent effective. In this case it seems that a very great deal got through.

Such 'diagnostic riddles' are not uncommon in the fluoride literature, but this particularly dramatic case has been widely quoted and broadcast by the antis in America. It is rumoured that it may have played a part in persuading the FDA to ban the use of fluoride in prenatal drugs in 1966. It is more likely, though, that there were other reasons and more serious ones, even if it is difficult to find out. The WHO 1970 report, for instance, doesn't even mention the ban and in Britain, doctors would only be likely to know of it if they were intimately acquainted with American anti-fluoridation literature.

Mongolism

The ban came suddenly and hastily in 1966, and the FDA gave a somewhat vague reason: 'there is neither substantial evidence of effectiveness nor a general recognition by qualified experts that such preparations are beneficial to tooth development in the foetus or in the prevention of dental caries in the offspring'.

Nobody explained how pregnant women might avoid fluoride in fluoridated drinking water. According to Dr Gunnar Bergström in Sweden, a FDA commissioner, Dr James L. Goddard, publicly stated a week after the FDA announcement that the reason for the ban wasn't just that fluoride tablets were not beneficial, but that there could also be a danger of mongolism and mental retardation.

The question of fluoride and mongolism has been an emotive one and the antis have, quite erroneously, tried to link it with thalidomide.

During the late 1950s and early 1960s Dr Ionel Rapaport at Wisconsin University did a number of surveys which showed that fluoridated areas had twice the rate of mongol births as unfluoridated ones. He had to publish his papers in France, and to this day the only versions in English have been translated by the antis and circulated unofficially.

Dr Rapaport, born in Rumania, graduated at the University of Paris in 1937 and was a veteran of World War II. He later went to America and specialized in the epidemiology of mental disorders.

When I first tried to contact him, I rather had the impression that he, and his work, was a figment of the fertile imagination of the lay antifluoridationists. I wrote three times to the University of Wisconsin and was told that they had never heard of him and had no records of anyone of that name connected with the university. Eventually I traced his name to the University's Institute of Psychiatry in Madison, and finally I had a note from the Library of the Department of Mental Hygiene of the New York State Psychiatric Institute, telling me that he had died of cancer in September 1972. His death was mentioned briefly in the *Journal of the American Medical Association* in January 1973. He was a quiet man, I have been told by someone who met him, and a man who appeared to have suffered 'many internal scars'.

His work caused an outcry, and despite its remarkable unavailability to English readers, is still being quoted today. I have no doubt that it will be quoted for a long time yet. Rarely can a work which, for all the ordinary American doctor knew, didn't exist, have caused so much nervousness in high quarters. Babies, after all, steal the show in any country, especially if someone has done something nasty to them.

Rapaport's first investigation was in 1956 in North and South Dakota, Illinois, and Wisconsin, where he found a higher proportion of children born in high fluoride areas to suffer from mongolism. This research was based upon a count of mongols being cared for in institutions. His methods were strongly criticized, especially by Dr A. L. Russell, chief of the Department of Epidemiology and Biometry at the National Institute of Dental Research – and even today Rapaport is still being criticized for this report by opponents who ignore later reports.

In 1959 he published another report – again in France – where he had followed to the letter all the advice given him by Russell. Russell had suggested that the new study should be drawn up strictly in terms of the residence of the mother and with reference to the known fluoride content of her community drinking water. This was done in Illinois only.

This second report again showed roughly twice as many mongol births at 2.6 ppm as at 0.2 ppm. In more detail, he found that in the 0.2 ppm area there were 34.15 mongols per 100,000 births, or 1 per 2,928. In the 2.6 area there were 71.59 mongols per 100,000, or 1 per 1,397. Moreover, the average age of the mothers of the mongol babies in the high fluoride areas were lower than normal.

These figures may sound convincing in themselves, but they are difficult to interpret since the rate of mongol births is much lower than that found elsewhere in the world. Usually it tends to vary round 1 in every 600-700, though there are considerable variations both up and down. By the age of 45 the rate may be as low as 1 per every 40 births.

The fact remains that the reaction to the paper was extraordinary. From 1963 onwards, according to Gunnar Bergström, official figures on mongolism were no longer published in Wisconsin, and it appeared to have become increasingly difficult to get hold of such figures in other states too. There was one report, again according to Bergström, in 1965 from the Cleveland Hospital Council, Ohio, which claimed that the rate of mongolism was now as high as 1 per every 300. There was another report from the Bartron Clinic in Watertown, South Dakota, which showed a rate of 1 per every 118, but this was not related to fluoride by the authors – though attempts have been made to do so by the antis – and it could not have been related to fluoride because of the confusing mixture of different levels of natural and artificial fluorides in the area and in the different wells. In any case the number of mongol births reported was very small. Between the years 1963 and 1973 the area has only had three mongol births.

The reaction to Rapaport extended to Britain. In 1958, W. T. C. Berry, Medical Officer at the Department of Health, published a survey of his own in the *American Journal of Mental Deficiency Research* – to the great surprise of some British doctors and dentists who, they have told me, would have expected such an important study to be published on home ground, if not actually by the HMSO. However, it appeared in an obscure American journal which few people until then had ever heard of.

This study included high fluoride areas in South Shields, West Hartlepool and Slough (0.7-2 ppm) and low fluoride control areas from Tynemouth, Carlisle, Gateshead, Stockton, High Wycombe, and Reading (all 0.2 or less). There were also figures for rural Essex County done by someone else, with varying levels from 0.2-4 ppm and areas with less than 0.25 ppm.

Berry claimed that his study showed that the rate of mongol births is much higher in Britain than in Rapaport's original study – which is certainly true – and that there was no relationship between the level of fluoride in the water and the incidence of mongolism.

In actual fact, if one adds up Berry's figures, the high areas show 64 mongols per 44,962 births or a rate of 1 per 702, and the low areas show 135 mongol births per 87,958 which is 1 per every 651 live births. Thus there are more mongols in the *low* area.

However, in Essex there were more mongols in the high areas: 1 per 1,102 compared with 1 per 1,514 in the low areas. The Essex part of the survey was done by the County Medical Officer of Health, Dr G. G. Stewart over a period of five years. This, of course, is totally contradictory. The paper is a short one and one rather has the impression that the bit with the Essex figures got stuck on to the end by mistake. That may be a biased impression, but the fact is that there have been vague allegations that there were unpublished figures which were startling.

Again, I do not know if it is true. Certainly the two sets of Essex figures seemed to have caused much more excitement than Berry's own, at least in some quarters. As the Department of Health does not publish separate figures for mongol births, it is not possible to delve any further into it, nor to relate this problem to fluoride levels in Britain today.

Several people – Professor Polya and Professor Steyn among them – have claimed that fluoride is mutagenic and even at low levels can produce chromosomal changes. Dr Roger Berry, of the Radiobiology Laboratory at the Churchill Hospital, Oxford, has said that 'there is no doubt that fluoride is mutagenic in high doses'.

Although there are many different theories about the cause of mongolism, it *is* just possible that the increase in fluoride levels in the body of the mother with age, could have something to do with it. Moreover, a report from the New York State Department of Mental Hygiene and the Division of Epidemiology at Columbia University School of Public Health in 1973 showed that the rate of mongol births in New York had increased almost a hundredfold between 1953 and 1970 – though one important factor was the increased lifespan of these babies with improved medical life-saving techniques. The US rate is now claimed to be four times that found in Hiroshima after the bomb.

There is a bit of circumstantial evidence of a possible connection – nothing 'proved', but intriguing suggestions for anyone who cared to investigate them. A report from the Department of Epidemiology and International Health and the Department of Pediatrics at the University of Washington in 1972 suggested that there is accelerated ageing in young mothers of mongol babies with, among other things, an increased prevalence of grey hair, and that this might be associated with either thyroid trouble or, alternatively abnormal glucose metabolism – though again both may independently be related to age.

In 1973, a report from the St Lawrence's Hospital, Caterham, Surrey, also discussed the finding that a significantly higher proportion of mothers who bore mongols when young had thyroid trouble. And in 1972, a report from the Department of Epidemiology and Health of McGill University in Canada showed that mothers of mongoloid children had a higher incidence of thyroid disorders, either hypo- or hyper-active. In fact, thyroid disease has long been suspected as a causative factor, but, again, there is no proof. Much of this must remain speculation, just as Rapaport's own findings really have not given us much to go by.

What is really odd about all this, and about Rapaport's publications especially, is the fact that there has *not* been a great upsurge in research into this possible connection with fluoride in the usual way in which scientific research is stimulated by controversial claims. After a.l, Frederick Stare's famous 'high-low' study is said to have caused no less than fifty research grants and funds to have materialized there and then. Why did not fifty different medical groups feel stung into investigating Rapaport's much more controversial claims about a much more awful disease? Did they just not know about it? Or was it simply that money was not forthcoming?

We shall never know. I have met American researchers doing vaguely controversial things in other controversial fields who – even while talking to me in London – would lower their voices while they told me that they did not want too much publicity, lest the US Public Health Service withdrew their grants. All grants from public funds in the US ultimately come from the US Public Health Service, an organization which, as we all know, is firmly committed to a pro-fluoride policy. Perhaps the only surprising thing is that Rapaport should have got away with it at all.

Fluoride and the thyroid gland

When talking about the thyroid gland, a bit of simple, old-fashioned chemistry might be helpful. There is a group of elements called *halogens*: fluorine, chlorine, bromine, and iodine. According to the so-called law of halogen displacement, any one of the four will displace the element with a higher atomic weight. Fluorine is the lightest of the four, and can displace any one of the other three.

There has been much speculation about fluorine displacing iodine in the body – and iodine is of course needed for the production of thyroid hormone. It is an old, old controversy, and it still hasn't been quite worked out. Is fluoride anti-thyroid and a goitrogen, or isn't it?

Bergström claims that the number of thyroid cancers in San Francisco has gone up five times since the introduction of fluoridation (though the original paper doesn't mention the word fluoride once), and that in Japan struma, or goitre, increased rapidly above a fluoride water level of 0.3 ppm. However, it is not as easy as all that.

Traditionally, endemic goitre is thought to depend on the presence or absence of iodine in the water, though it is known and accepted that goitre *can* occur in high iodine areas and conversely, there may be an absence of goitre in low iodine areas. Iodine itself is in many ways as complicated and controversial as fluorine.

The WHO report states that 'a tendentious interpretation of certain medical and biological observations led to the erroneous theory of iodine-fluorine antagonism,' and also points out that this led chemists to synthesize inorganic and organic antithyroid fluoride drugs for medical treatment, though of these drugs not a single one is still on the market. Apparently they were not very effective. The report also states categorically that 'consumption of drinking water containing fluoride, either naturally or artificially, does not impair thyroid function', and even that 'the consumption throughout life of water containing 6 or 7 ppm fluoride does not affect the thyroid'.

That, however, is not the whole picture.

For a start it seems that there are so many anti-thyroid substances around in the civilized world nowadays that it would no longer make sense to compare the situation in the West with the kind of 'simple' surveys in old-fashioned endemic areas on which previous hypotheses were based. This is quite apart from the fact that thyroid function is mixed up one way or another with other physiological functions, such as the sex hormones and menstrual disorders in women, and even with mental illness and depression. Thus lithium – one of the candidates for being added to drinking water in the future – is antithyroid and may interfere with the iodine metabolism, though lithium is used in combination with conventional drugs to treat manic-depressive illness.

Dr B. J. Mayer, Department of Physiology at the University of Pretoria Medical School, and Dr E. U. Beiler and Dr C. R. Jansen, both of the Life Sciences Division of the Atomic Energy Board, Pretoria, recently pointed out in the *South African Medical Journal* that the incidence of hypothyroidism in white patients was exceptionally high. They distinguished between old-fashioned endemic goitres occurring in regions with iodine deficiency, and 'sporadic goitre' occurring outside such regions, and claimed that there has been an increase in both naturally occurring goitrogens in food and other antithyroid agents in pharmaceutical drugs, and this seems to have happened over the past 10-20 years.

They do not mention fluoride. But the British pro-fluoridation lobby were much upset in 1972 when two young doctors from Guy's Hospital, London, P. R. Powell-Jackson and Thomas K. Day, showed after a survey in Nepal, that fluoride should still be taken seriously. This was published in *The Lancet*. The results showed that where iodine is not plentiful, the hardness of the water may be more important in determining goitre prevalence. Thus in Britain hard waters have also been associated with goitre, though it may be magnesium rather than the calcium in such water that causes the trouble. However, so does fluoride, and fluoride-rich waters have traditionally been called 'goitre water'. In general, the data showed there was most goitre either in areas connected with a high fluoride level, or, more so, in hard water also with high fluoride levels. The same two writers have also objected to the apparent endorsement by *The Lancet* of the use of fluoride toothpaste – a method of medication which they regard as 'imprecise' and 'irregular'.

'Perhaps in our enthusiasm for preventive dentistry we have forgotten that the level of fluoride intake which is effective in preventing caries is dangerously close to the level which produces harmful effects.'

The problem of softness or hardness of the water and goitre has also been taken up by Dr Margaret D. Crawford at the London School of Hygiene and Tropical Medicine, again in *The Lancet.*

Commenting on the paper by Day and Powell-Jackson, she said that it 'raised a problem concerning fluoridation of water-supplies in this country which has not received sufficient attention, that is, the relationship between fluoride and other ions present in drinking water, in particular iodine. A Medical Research Council memorandum reported that in some areas even moderate concentrations of fluorides in drinking water could block iodine absorption. It is also known that iodine concentrations are lower in soft water than in hard waters; in a study of trace elements in drinking-water, it was estimated that, on average, there was five times as much iodine in hard as in soft waters. Naturally soft waters, however, have little or no fluoride and the iodine is therefore all available for absorption. If fluoride is added to soft waters this will not be so and a proportion of the population may come to have sub-optimal iodine intake. The effects might be subtle and slow to develop and would certainly not be picked up by the crude screening used at present'.

Traditionally it is known that mottled teeth and struma often occur together, though a high intake of fish may prevent the struma. All this has been thoroughly written up by classical writers in the field, such as the Dane Kaj Roholm, whose book, *Fluorine Intoxication, A Clinical-Hygienic Study,* written in 1937, is still today a kind of fluoride Bible, and by T. Gordonoff and W. Minder in Switzerland, two well-known researchers in this field (though the WHO report practically ignored them).

'Fluoride,' they say, 'is by no means a harmless substance.' The toxicity of fluorine is related to calcium intake and to calcium metabolism generally, for the only therapy in acute and chronic fluorine poisoning consists of administering calcium. In any event, 'we have here an element of narrow therapeutic range' and one which 'easily lends itself to overdosage'.

Their writing is full of references which tell us that toxication with fluorine has been observed in Iceland after volcanic eruption, that mottled teeth occur in Iceland, that both mottled teeth and struma is known in Utrecht in Holland in connection with the drinking water, and in both animals and people. Trouble of this nature from fluoride does not only occur in faraway places like India or the Middle East. It is here, on our very doorstep – and yet the public is never told about it.

Cancer and Paradoxical Effects

At the famous meeting of state dental directors in 1951, Dr. Glover Johns, Associate Director of the Division of Dental Health of the State Department of Health, Austin, made the following remark:

'There was a rumour that this research project indicated that fluoridation of water supplies causes cancer. That has knocked the pins from under us. We don't know how to combat it . . . the University of Texas is now sorry it happened and doesn't know how to stop the rumour.'

The University of Texas can't have been all that sorry. Fourteen years later the culprits guilty of starting this obnoxious rumour were still at it.

True, they published in fairly obscure journals such as the *Proceedings of the Society for Experimental Biology and Medicine* – but for people who are publicly abused by their opponents even before they publish, publishing in even the least-known of journals would not prevent them from becoming famous, nor would it prevent the 'rumour' from spreading.

Between the early 50s and still in 1965, Alfred Taylor and Nell Carmichael Taylor of Texas University published a series of papers on the effect of sodium fluoride on tumour growth in mice. They found first that fluoride reduces the lifespan of mice, and later that cancer growth was stimulated by fluoride. This observation held true regardless of whether the fluoride was added to the drinking water, introduced by injection or added to suspensions of cancer tissue prior to inoculation.

Drs Roger J. Berry and W. Trillwood of the Radiobiology Laboratory, Churchill Hospital, Oxford, produced similar claims in 1963, though it seems that Dr Berry has done no work in this field since, because he found the political climate too hot. His work caused an angry reaction and letters in the *British Medical Journal* – one from Professor G. Neil Jenkins who tried to re-interpret the findings to suggest that they actually supported the safety of fluoridation, and another from C. G. Learoyd who suggested that the research 'might conceivably be of interest to those who have patients with cancer of the cervix or mice with rheumatism! ' Learoyd also attacked *The Times* for having written up this research and said that the newspaper 'never dropped a hint to their horrified readers that every article in human diet contains this invaluable trace element and anyone eating a couple of grilled kidneys or drinking tea is in imminent danger of killing his cells'.

These remarks are not far from the classical pro-fluoridation sentence which starts, 'when people discover they don't die like rats . . .'. Such irresponsible use of words would in itself cause an outcry, if used in any other medical context.

However, there has been serious interest in a possible connection between fluoride and cancer elsewhere. In Japan, cancer has been linked with a food called 'miso', a fermented product made from rice and soya-beans, and Soviet researchers have looked at high cancer rates in the vicinity of aluminium factories.

Albert Schatz has claimed that the death rate from

leukaemia and all types of cancer has risen in Birmingham (England) since the introduction of fluoridation. Cancer has also been linked with phosphate fertilizers and even with toothpaste. (Nader, moreover, has alleged in the US that fluoride toothpastes have been shown to cause stains of the teeth, and he wants warnings to that effect put on the labels.)

One paradoxical point put forward by Taylor is that while small doses of fluoride produce tumours, large doses appear to inhibit cancer development. This is interesting in the light of new research in America – by Dr Alfred S. Ketcham, Clinical Director and Chief of Surgery at the National Cancer Institute – and by Dr Douglas Thornes, Professor of Experimental Medicine at the Royal College of Surgeons in Dublin – that large doses of sodium fluoride can be used successfully to retard or check tumour growth, in combination with more conventional treatment. This work is a spin-off from the use of fluorides as anticoagulants in the treatment of heart disease. What is really intriguing here is the difference between large and small doses.

William Lijinsky, of the Oak Ridge National Laboratory's biology division recently stated, '. . . our knowledge of the working of chemical carcinogens, such as it is, shows that they act optimally when administered as frequent small doses over a long period, rather than as large scale doses'. He did not refer to fluoride, but the problem of the small dose is one which has been studied by Albert Schatz who, apart from his interest in dentistry, is also a cancer researcher.

Schatz has collected a lot of independent references from different authors who have noticed the unexpected appearance of what is called 'paradoxical effects' in fluoride research, a phenomenon which is better known in soil chemistry and other fields than in medicine.

The effect means simply that a higher dose does not necessarily produce a more pronounced effect. Thus, for instance, at 1 ppm there may actually be increased enamel solubility (though fluoride is supposed to lower this), while at 10 ppm the effect is reversed, only to reappear at 100 ppm. Or the enamel may be less soluble at 200 ppm than at 100 ppm or at 400 ppm.

This is something which has puzzled researchers for a long time, though there does not seem to have been any efforts on the part of the different researchers to get together and try and work it out. On the contrary, according to Schatz, they often appear ignorant that others have met the same unexpected mystery. For immediate medical purposes it may not be important, but from the point of view of a long-term influence that could conceivably end with cancer, it *could* be important. What it means is that one should be on guard against the possibility that in some circumstances a small dose may have a worse effect than a large one. It also means that the whole basis for fluoridation, namely that below a certain level fluoride is harmless, is not at all certain.

When considering cancer, it is always important to remember that other factors which occur at the same time, may be equally important. Thus a survey in Canada which showed increased mortality from lung cancer in the proximity of steel mills in the city of Hamilton, Ontario, speculated that the cancer could be attributed to other cancer-producing agents emanating from the factories, namely nickel, carbonate, chromium, beryllium, and silicates. However, it is thought that beryllium combined with fluoride has a much greater carcinogenic potency than either element alone, and thus synergistic effects play a part as well. Something similar may happen with asbestos, long linked with cancer, which also contains fluoride.

The fact that some agents may be dangerous in high doses and yet be 'natural' in small ones, as often argued by the fluoride promoters, also does not alter the picture. Dr Ludwick Gross, of the Cancer Research Unit at the Veterans Administration Hospital in Bronx, NY, and an anti-fluoridation man, has pointed out that 'humans and animals carry small quantities of many poisonous chemicals and potentially deadly bacteria, moulds and viruses.

They can be called 'natural' components or companions, because they form what is part of the hazards of life. It would be, however, a rather far-fetched assumption to feel that they are (therefore) essential for optimal health.'

Allergy

There are a number of other diseases and conditions which have, from time to time, been linked with fluoride, though not as fully as those mentioned here. One important – though difficult – one is allergy, which has been researched by Dr. George Waldbott, editor of *Fluoride*. Waldbott is a well-known opponent of fluoridation, and the authors of the famous 1962 issue of the *Journal of the American Dental Association* replied to his claims by lumping him together with various cranks and with such organizations as the Ku Klux Klan, and the John Birch Society (who at the time had decided that fluoridation was a Communist plot to poison America's water supply).

He has been attacked and ridiculed on other occasions, among others by Dr Francis Bull, who at one debate in Milwaukee remarked about him that 'it's astounding that we have to get a doctor from outside the state to tell us that people here are walking around half dead'. Dr Waldbott's own answer to such attacks is an excellent journal, in the circumstances, and one which may one day turn out to be a goldmine of information for those who wish seriously to study all aspects of fluoridation.

Organofluorides

One last word is necessary about the organofluorides. These are carbon compounds, distinct from the inorganic or mineral compounds discussed so far. Some can be synthesized by food plants exposed to fluoride air-pollution and produce symptoms quite different from those associated with inorganic fluoride, such as blood sugar irregularities, convulsive seizures and respiratory failure which may result from ingestion of vegetation contaminated by insecticides.

On the whole, the main concern is from such organic

fluorine gas propellants, for instance, which are also called Freons, and which are used in aerosol dispensers for cosmetic, household or pharmaceutical purposes, or similar compounds released at high temperature from some types of 'plastics', such as Teflon which is commonly used for frying pans. Fumes produced by heating Teflon pans to temperatures slightly over 500°C have caused death in small pet birds within 20-30 minutes.

Another compound belonging to this group is the drug fenfluramine, used as an appetite-depressant for slimmers, and yet another group are fluorinated anaesthetics, such as halothane.

The problems caused by the latter are well-known to the medical profession and arguments and counter-arguments have been appearing in leading journals for some time.

The main danger to patients from anaesthetics of this kind is death from heart and kidney failure. It seems that there may be a conversion from organic to inorganic forms in the body which the kidney can't cope with.

A major researcher in this field is Willard S. Harris, from the Department of Medicine at the Abraham Lincoln School of Medicine, at the University of Illinois, Chicago. Researchers at Rochester University have also written about the problem.

It is equally serious, though more hotly disputed, as a danger to operating room personnel, from inhalation, and it has been connected with foetal abnormalities and spontaneous abortion in female anaesthetists. It has been claimed that the rate of spontaneous abortion among personnel doubles subsequent to their employment in the operating room. Liver trouble, headaches, fatigue, loss of memory and disturbances of the central nervous system are all frequently mentioned.

Such fluorinated hydrocarbons or Freons in aerosol propellants and even in glue, are now looked upon as potential cardiac toxins. During the past five years 140 'healthy youths' have unexpectedly died minutes after trying to become intoxicated by inhaling the gas from aerosol propellants. Dr Harris thinks that conditions such as asthma, heart failure, coronary artery disease may increase susceptibility to the toxic effects of these gases.

They have been in use since World War II, and in America more than two billion aerosol packages are marketed every year and the number is growing – but practically no research has been done on the long-term effect of regular daily exposure.

VI

The 'Wild-cat' Element: 'I want it for my child'

'I want it for my child!' said a girl journalist angrily once when the word fluoride cropped up in the conversation. 'These fanatics have no right to keep it out of the water.'

At that very moment the dental magazine, *The Probe,* carried an amazing and perhaps somewhat extreme series of anti-fluoridation articles by a dentist in Dunelm called F. R. Bertrand, in which he described fluorine as the 'wild-cat element. . . . because it can be difficult to forecast how it will react'. He even tried to show that fluoride could be linked with mental disturbances and with the troubles in Northern Ireland (quite erroneously, I'm sure), and in due course received his measure of ridicule in the correspondence which followed.

Perhaps he deserved some of it. He would not have convinced the girl journalist – though it is one of the delights of the British way of life that such quirky happenings surprise you where you least expect to find them. I do not remember seeing an anti-fluoridation article in any other dental publication. The author even seemed to be against chlorine.

Even so, there was nowhere in London in any easily accessible journal, magazine, or library, where the girl journalist could have discovered exactly how much fluoride her baby was already getting. She had talked to the dental experts and was obviously under the impression that her baby wasn't getting any fluoride at all. She would have been surprised at the advice which appeared in the *1939 US Yearbook of Agriculture* which stated, '. . . it is especially important that fluorine be avoided during the period of tooth formation, that is, from birth to 12 years.' '. . . this dental disease (mottling) is always found when water containing as little as 1 part per million is used continuously during the period of formation of the permanent teeth'.

In fact, Dean is said to have admitted that he found mottling in areas with as little as 0.3 ppm in some places, but she wouldn't have known that either. Nor would she have known that in Colorado Springs (with 2.5 ppm), according to an official Public Health report in 1957, dentists and paediatricians had 'recommended since 1935 that parents provide their children with low-fluoride water during the development of the permanent teeth as a preventive measure against this disease (mottling). The local dairies have cooperated by supplying low-fluoride (0.2 ppm) bottled water from their private wells, delivering it to the homes with the milk'. Despite this, a former president of the Detroit District Dental Society, Alfred E. Seyler, is claimed to have said that 'people in some parts of the country – Colorado Springs for instance – drink a natural concentration of *20* parts per million of fluorine without ill effects'.

The early efforts with regard to fluoride was to *remove* it from the environment, and it is significant that agricultural authorities have continued to worry about fluoride all along, regardless of the dentists, and that the best surveys and most comprehensive and trustworthy reports have come from the scientists concerned with the environment.

The best and most up-to-date is *Environmental Fluoride* by John Marier and Dyson Rose at the Division of Biology, National Research Council of Canada.

'During the past 100 years,' they say, 'man has been responsible for steadily increasing the distribution and availability of fluoride in his environment', and they point out that any process that involves new materials taken from the earth's crust and subjected to heating at high temperatures will liberate fluorides. Some fifty different types of industries are now involved in fluoride pollution, according to Nader.

Waldbott has mentioned that fluorine compounds are employed in the production of metals such as aluminium, steel, uranium, beryllium, in manufacturing bricks, pottery, enamel, plastics, and in processes involving refrigerants, lubricants, expellants. They are used in the missile programmes, and are present in many drugs, in steroids, tranquillizers, anti-cancer drugs, anaesthetics, and as contaminants in calcium preparations. They are, in America, also added to vitamin preparations, and of course fluorides are found in toothpaste, in fluoride tablets, and in the drinking water. They are not hard to come by, and even fluoride toothpastes are slowly polluting the environment, according to Albert Schatz. Toothpastes alone, Schatz thinks, may be adding 116,000 pounds of fluoride to the environment every year. But from the pollution point of view, the various fluorine gases are, if anything, even more important.

In 1967, the American Association for the Advancement of Science listed fluoride as the third most serious air pollutant in a group headed by sulphur dioxide and ozone. In 1966 the National Conference on 'Pollution and Our Environment' in Montreal decided that 'prolonged exposure to ambient air concentrations of less than 1 part per thousand million parts of air volume of fluoride may create a hazard . . . in this respect fluorides are 100 times more toxic than sulphur dioxide'.

Marier and Rose claims that one steel operation in America returns its fluoride wastes back to earth in 'burial plots' similar to those used to dispose of radioactive wastes. They also quote a 1962 report on air and water pollution which examined a US Atomic Energy Commission plant producing uranium and thorium metal, in which it was stated 'The problems of waste disposal . . . are many and varied. The biggest single problem is fluorides'.

There is obviously a lot of money involved here, and though by far the biggest financial problem is the cost of disposal, there is also profit to be made. In 1951, *Chemical Week* is quoted as noting that 'the market potential has fluoride-producers goggle-eyed,' and in 1964 that by recovering fluorine as fluosilic acid, producers would make a profit on what would otherwise be objectionable waste products. In 1964 in Toronto, fluosilic acid, produced at 3-5 dollars per ton, was selling for 45 dollars per ton. Toronto used 30 tons of hydrofluorosilic acid *per week* for water fluoridation.

In agriculture and animal feeding there never was any doubt that fluoride is undesirable.

'The US Department of Agriculture warns the farmers of America against the contact of any of their stock with fluorides in any form. Does this mean that the cattle of America are more important than the People?' asked one writer in *American Druggist* in 1971, Mr. H. L. Thomson of the Thomson Drug Company, Arkansas.

Already in 1955 – according to a leaflet issued in 1966 by the Physicians' Committee of the Greater Detroit Citizens Opposed to Fluoridation – a judgement was issued against the Reynolds Metals Company's plant in Portland, Oregon, because of serious injury from fluoride-contaminated air to the health of an Oregon farmer's family. This was the first case in the US in which harm to human health had been proved, despite the fact that seven chemical and aluminium corporations joined forces with Reynolds' in an unsuccessful attempt to obtain a reversal of the decision. All along, however, there has been much litigation over damage to farm animals. Thus, in and before 1950, Alcoa's Washington plant had dumped between 1,000 and 7,000 pounds of fluoride per month into the Columbia River with resultant injury and death to cattle in the area. Yet instead of outlawing disposal into waterways of all fluoride wastes, the US government in 1942 set 1 ppm as the top allowable limit in water used for drinking purposes, in 1946 the maximum was raised to 1.5 ppm, and in 1961 to 2.4 ppm, despite no new evidence of safety. The Detroit physicians behind this leaflet complain bitterly that doctors have not been properly informed about fluoride pollution.

'Fluoride damage to animal and plant life and to

humans has been widespread in the vicinity of phosphate-fertilizer factories in Tampa, Florida, and in the Tennessee Valley area. With industrialization of the West, the nuisance spread to western states, especially Utah, Montana, and California, as well as Oregon and Washington. Near fertilizer factories it is not unusual to see cattle crawling across pastures on their knees.'

Fluoride-induced lameness in cattle, and also other domestic animals, had been known since before the war in Denmark, Germany, Italy, Switzerland, and Belgium, and was what first aroused the interest of Kaj Roholm. Yet today it seems, in North America at least, that the subject of fluorosis isn't always taught even in veterinary schools. Dr Waldbott was involved in a case in Sheerbrooke, a small community in Ontario, where both cattle and people were affected. Two young vets, just graduated, had just then set up a practice in the area and went about trying to discover what the mysterious disease was. They were baffled that the only consistent clinical symptom in all cases was a very high fluoride level in the urine among the cattle. It appeared that they had not been taught to look for this symptom, and did not at first connect it with the fertilizer plant built in the area by the Electric Reduction Company Ltd, nor with fluoride damage to crops.

There are many stories like this one in American anti-fluoridation literature, though it is difficult to obtain independent information. The symptoms reported, however, are identical to those discussed at a CIBA Foundation Symposium on carbon-fluorine compounds and published by Elsevier in Holland in 1972. There can be no doubt that this is a serious problem.

In 1970, a US Department of Agriculture handbook included a review by a Robert J. Lillie, in which he stated that:

'Airborne fluorides have caused more worldwide damage to domestic animals than any other air pollutant', and 'whenever domestic animals exhibited fluorosis, several cases of human fluorosis were reported . . . man is much more sensitive than domestic animals to F intoxication.'

At the first European Congress on 'The Influence of Air Pollution on Plants and Animals' at Wageningen in April 1968, it was stated that airborne fluorides had destroyed 400,000 hectares of European forests. It was also claimed that in Norway, fluoride injuries to coniferous forests could occur at a distance of 32 kilometres from the emitting source. It was stated at a symposium in Barcelona in 1969 that a German survey of areas with coal-burning industries revealed that rainwater could contain up to 14 ppm fluoride, or 88 times the level in the control areas. These levels correlated directly with those found in vegetation. Moreover, grazing near a fluoride-emitting factory could result in a 50-100 increase in intake. In one German survey, forage was shown to supply more than 90 percent of the fluoride ingested by cattle.

The damage done to plants and vegetation by fluorides is complicated by wide variations in the way different species react and the fact that some of them may accumulate very high levels without apparent damage, while others suffer from low levels. Many different types of fluorine compounds are also involved. Moreover, plants may take up fluoride through the roots as well as from the air. Some forms of edible vegetation have the capacity to produce fluoroacetate and fluorocitrate, and this may be stimulated by exposure to environmental levels of inorganic fluoride. Lettuce is one of these. Marier remarks, '. . . the toxicological significance of these findings awaits thorough investigation. Fluorocitrate is known to be extremely toxic to mammals, and it appears probably that the presence of such organofluorides will markedly alter the pattern of "toxicity-symptomology" accruing from ingestion of fluoride-polluted forage plants and of edible vegetation consumed by humans'. He adds that no data seem to be available concerning the limits of fluorocitrate in vegetation. Such limits, he thinks, ought to be next to 'nil' in vegetation destined for ingestion.

Again, as in so many other fields of fluoride science, the public hasn't been told, the work hasn't been done, and the public hasn't been told that the work hasn't been done. Fluoride pollution is not supposed to exist.

Unexpected problems may occur. In 1972, a Swedish dentist, Dr Sven Bramstaang, claimed that the occurrence of a new and mysterious skin rash among bathers along the coast of Sweden and Denmark was caused by fluoride waste products pumped out into the sea by phosphate manufacturers. He claimed that although these might not be dangerous in themselves, they appeared at certain temperatures to combine with acids in other industrial wastes and turn into the highly dangerous hydrogen fluoride. The skin rash – which consisted of ulcerating wounds ('not unlike those found in lepers') – also appeared especially in children who had been given fluoride tablets at school. Dr Bramstaang thinks something similar could happen internally in patients with stomach ulcers (a disease which is on the increase among young children). The authorities tried to claim that the rash was caused by leaking mustard gas bombs off the coast of Bornholm in the Baltic, but these are some 200 miles from the area where the bathers suffered (and incidentally only five miles from a beach in Bornholm where my family owns a cottage, and no skin rash had appeared among the Bornholm bathers).

This one remains to be proved, at least at the time of writing. It was only one man's claim against the determination of the authorities to deny all connection with fluoride.

Albert Schatz has been particularly concerned about the fluoride hazards from phosphate fertilizers, and the effect on fluoride levels in food, but such 'extremists' are not the only people who from time to time find themselves utterly surprised at what is going on. Thus Sir Joseph Hutchinson, Drapers' Professor of Agriculture at the University of Cambridge wrote in *The Advancement of Science* in 1967 that 'on the same day my morning paper had an article on the unreasoning opposition that had been met by those who want to put fluoride in water supplies, my mail included a press release from the Ministry of Agriculture reporting the results of an eight-year experiment on methods of ameliorating the damage to dairy cows from fluoride deposited from certain industrial smokes on their pasture.

'Fluoride in the proper dosage, like chlorinated hydrocarbons in their proper place, can be used for the improvement of human welfare. Out of place, it is a more serious pollutant than chlorinated hydrocarbons in that it is indestructible, and we should take care that we do not have to learn the lesson of persistence and accumulation a second time.'

Not really an extreme remark. Schatz reviewed the fertilizer problem in the March-April 1972 issue of *Compost Science*. The application of 1,000 pounds of superphosphate to an acre adds appromixately 17.5 pounds of fluorine. That increases by about 7.5 ppm the fluorine content of the soil to plough depth. The same amount of rock phosphate adds about twice as much fluorine. In 1970, he points out, Soviet soil scientists reported that regular application of these phosphates during 35 years increased the content of fluorine in the soil and in the plants, and this resulted in a decrease of productivity, especially in maize. In 1946, similar observations were made by the US Department of Agriculture, and in 1943, researchers at the University of Wisconsin remarked that 'when phosphate fertilization is carried on over many years, very considerable quantities of highly toxic fluorine will have been added to the soil. The fact remains that dangerously high concentrations of fluorine are possible in the drainage water from fields bountifully supplied with phosphate fertilizers.' 'This,' they added, 'raises the question whether our present system of soil fertilization with fluorine-carrying phosphates may lead to a contamination of drinking waters to a point where they may become dangerous to human health.'

In Japan, an alarming increase of fluorine in some common foods grown in the district of Aichi was found in 1965. These included pumpkin, green tea, watermelon and lotus rhizome. The fluorine content of the latter increased by 976 percent. In some cases, the daily human intake of fluoride in the the diet was 11 mg. The same researchers also found a correlation between fluorine content of rice and the death rates from gastric cancer, and between cancer and 'miso'.

In 1964 the US Department of Agriculture, in an official publication, suggested the following, and peculiar 'compromise':

'. . . for . . . animal feed supplement . . . the primary objective is the production of fluorine-free phosphate. But for fertilizer purposes, the objective is to make the phosphorus agronomically available without regard to fluorine.' With such somersaulting by the authorities, no wonder farmers in high-fluoride areas are in trouble!

In November 1972, Professor D. Jackson from Leeds University, in defence of adding fluoride to the water, said that fluoride added to food would be too haphazard. It is an extraordinary remark, not only in view of the totally haphazard occurrence of fluoride from industrial sources in food, but also when one considers the fact that as far as British law is concerned, water is not a 'food'. Fluoride additives are allowed in certain types of flour and in baking powder, partly because they are natural contaminants of the raw material used, but otherwise you can't add it to food. When I rang up the Ministry of Agriculture to check on this point, I discovered that the press officer had never even heard about fluoride being present in food, naturally or artificially. He thought it was just something which dentists added to the water to preserve children's teeth. He said he would first have to consult with the Department's chemists, and returned to tell me that fluorine and fluoride were not the same thing. Fluoride is what you add to the water to prevent tooth decay, fluorine is what might just occasionally make cows a bit unwell. That is not to say that the Government chemists got it wrong – I would normally have expected that at least agricultural chemists knew about fluorides even if nobody else did – but one cannot expect the ordinary press to be well-informed, if their sources of information are so totally in the dark.

The food question has been investigated by Marier, who, in 1966 reported in the *Journal of Food Science* an intake of up to 5 mg a day from all sources in healthy individuals under normal indoor vocational conditions; labourers exposed to outdoor summer conditions would be expected to ingest still more. His research also showed that the use of fluoridated water in food processing will cause a significant increase in the fluoride content of food and beverages.

Waldbott, in the June 1963 issue of the *American Journal of Clinical Nutrition* reviewed the literature on fluoride and food in an excellent report which, like Marier's reports, ought to be read by anyone concerned with the medical field of nutrition. It is unlikely at present that they would be able to get such comprehensive information from any other sources.

Most foods contain fluoride at least in minute amounts. Some foods concentrate additional fluoride from boiling, processing, or contamination, but levels may vary widely even between samples of the same kind of food.

Naturally occurring fluoride tends to be calcium bound, but other types may be taken up by plants from the air and through the soil. Even then the levels depend on the locality, the soil, the weather and the climate, the distance from contaminating sources, on whether the plant was fertilized or sprayed, and finally on how the food is processed.

There is relatively little fluoride in fruit and seed – though an apple sprayed with fluoride-containing insecticide may provide as much as 1 mg of fluoride.

Vegetables tend to have low levels early in the season, with rising levels after June. The highest intake occurs in dry periods, the lowest after rain. External parts of fruits and vegetables contain more fluoride than internal. Potato skins or peel may have five times more fluoride than the inside of the potato. Leafy vegetables contain more fluoride than other types. Lettuce, spinach, parsley, cabbage and kale have high levels. Tea tops the list and may provide as much as half a milligram per cup.

Fish and sea foods, including sea salt, contain high levels, and so does kidney, liver, and heart muscle.

When food is boiled in fluoride-containing water, its fluoride concentration increases. Cauliflower and cabbage absorb more fluoride than beets and carrots. Vegetables cooked in a saucepan absorb more than those cooked in a pressure cooker. Teflon pans emit fluoride at high temperatures (and indeed manufacturers often warn against overheating them, though they do not explain why).

Beer brewed with fluoridated water may contain 1.2 ppm. In America, some breweries in fluoridated areas use non-fluoridated water for brewing to be on the safe side. In Canada, however, ginger ale made with fluoridated water contains 0.77 ppm as compared with an unfluoridated level of only 0.02 ppm.

Processing methods make a considerable difference to the fluoride level of all manufactured foods, and dyes and detergents contribute to higher levels. In Canada, unfluoridated tomato soup contains 0.04 ppm but fluoridated tomato soup as much as 0.38 ppm.

Drugs like tranquillizers and steroids also often contain fluorinated compounds and may provide a daily intake of up to 1 mg. if taken habitually. In America people may furthermore ingest considerable quantities of fluoride in vitamin tablets.

Fluoride indeed is everywhere – only the public hasn't been told. To the fluoride promoter the very fact that fluoride is everywhere, is proof of its safety. It is a neat and handy argument which can be used about a great many other substances.

It has already been used about some of the others. Aluminium is one of them, and deserves mention because it is sometimes confused with fluoride.

The question of the safety or otherwise of aluminium cooking utensils crops up from time to time in fluoride and anti-fluoride literature. A Dr Leo Spira, who did his

Ph.D. in physiology at the Middlesex Hospital Medical School, London, was severely ridiculed in the November 1962 issue of the *Journal of the American Dental Association* for claiming that his research on animals had shown that both fluoridated drinking water and aluminium cooking utensils were dangerous. It seems his original research was done as long ago as 1928. To quote the experts from the American Dental Association:

'His statement that "fluorine was the causative agent . . . on account of its contaminating food prepared in aluminium cooking utensils" represents a most remarkable chemical concept. Unless Dr Spira has found the Philosopher's stone so long sought by alchemists, it is difficult to understand how the aluminium is transformed into fluorine. He may be confused into thinking that fluorides accompany the aluminium since fluorides are employed in the preparation of aluminium metal from its ore! '

He also, however, asserted that aluminium itself was toxic. That this should be so has been passionately denied by fluoride promoters, but as late as 1972 a writer in *The Lancet*, G. M. Berlyne of the Negev Central Hospital and the University of Negev's Faculty of Natural Sciences in Israel, wrote:

'The question, is aluminium toxic? – has never been answered satisfactorily.' Berlyne and his co-workers feel that aluminium salts should not be regarded as inert, and link them with a particular type of osteoporosis seen in Newcastle and accompanied by a high aluminium content in bone. They also link it with phosphate depletion. These points were taken up in the *British Medical Journal* the same year by Dr Sidney Shaw of the Department of Haematology at the Charing Cross Hospital, who said:

'I would stress that it is no longer safe to regard aluminium and aluminium salts as at present used as inert. There are possible dangers from the wide-spread practice of using aluminium containers for cooking and canning foods as well as in various medicaments . . .'

In fact, American anti-fluoridation groups have produced at least two leaflets with dire warnings of the dreadful consequences of using aluminium for food preparations, and – as would surprise no one by now – hint at yet another secret and unpublished investigation.

This claim came in 1954 from one of the characters ridiculed in the November 1962 issue of the *Journal of the American Dental Association*, a certain (and now deceased) Dr Charles T. Betts, Ohio, who was apparently against both vaccination and pasteurization but mainly concerned with 'the alleged poisonous characteristics of aluminium cooking utensils'. However, he quoted someone who appears to have been more 'respectable': Edward M. Averill, who did a report – Docket 540 – for the Federal Trade Commission after an alleged investigation into possible dangers from aluminium. This document, it seems, was still confidential in 1954 when Betts began making a fuss about it, though Betts claimed that Averill was appointed to open the case back in 1920. It took five years, 158 witnesses and 1,000 exhibits; 4,700 typewritten pages of testimony were taken. Betts apparently got hold of a copy of the report and decided to publish it, but was stopped by the Federal Trade Commission.

That at any rate is how the story goes. Needless to say, the *Journal of the American Dental Association* did not mention the secret report, but in 1965 it cropped up again, this time in the *National Police Gazette*, echoing quotes which seem to have originated from Betts. It would be pointless of course to try and evaluate claims from such vague sources and from a report which – if it exists – still has not been published. At least it seems that the American police have been genuinely concerned with food poisoning traced to aluminium utensils and containers, and it is obviously a subject which bothers medical researchers from time to time.

Needless to say, the powers-that-be have refuted all such claims. The Kettering Laboratory, originally set up by the US industry to investigate chemical hazards in industry and an organization that does not publish its findings unless allowed to do so by the sponsor in question, has been particularly keen to show aluminium to be absolutely safe. We should not perhaps be surprised at this, since the Kettering Laboratory is also the place where the alleged 10,000 (or 16,000 or 28,000) papers which prove fluoridation absolutely safe, are said to be lodged. In 1957 and again in 1967, with the Department of Preventive Medicine and Industrial Health at the University of Cincinnati, the Kettering workers published a report which more or less proved aluminium absolutely safe, too.

The evidence apparently consists of proof in the version of 1,500 books and articles 'sifted' by members of the Laboratory in 20 different countries. And the evidence sounds strangely familiar:

All vegetation contains some aluminium. Aluminium occurs in 'natural waters' (although usually at very low levels). Aluminium is 'highly suited for the cooking, processing and storing of foods'. It is a 'superior' material for the storage of water, vinegar, cold juices, vegetables, fats, oils, fish, milk and meat, and 'aluminium makes a significant contribution to the safety of the food supply'. One is not far from gaining the impression that if food has gone bad, you only need to store it in an aluminium container and it'll turn good again.

'Aluminium is distributed so abundantly in the earth's crust that it is only natural that it be present in the air breathed by man and in the food eaten by him.' 'Alum is widely used as an astringent and mordant to keep the firmness and colour of preserved fruit, to clarify syrup in the refining of sugar, to remove undesirable particles from beer in brewing, and to settle impurities out of public water supplies.' 'Baking powders containing aluminium compounds are no more harmful than other baking powders.' There is also talk of steady levels in the diet, as well as physiological barriers to the absorption of aluminium in the body which further proves its safety.

Of course there has been a bit of lung damage from inhalation, but this has not been found to be associated with disability. It is also acknowledged that there is a thing which physicians call 'aluminium lung', but this, it seems, occurs only in plants. No, what the authors must mean is that it occurs in *workers* in aluminium plants. Nevertheless aluminium is everywhere and that is proof enough that there is no cause to worry. God, in fact, has given us aluminium, just as he has given us fluoride, and we should be grateful. You could take this leaflet and substitute the word 'aluminium' with the word 'fluoride' throughout, and one would hardly notice the difference.

Now, of course, you also get fluoride in baking powder, and these baking powders are exempt from the usual legal demands about fluoride levels in food, and so is water, naturally. The aluminium bit is perhaps a side-step, and it is perhaps inevitable that people who are against vaccination, should also be against fluoride as well as aluminium. What is so fascinating for outsiders like myself is the extraordinary similarity in all official safety assurances. And of course, lurking in the background is the aluminium industry. *They*, at least, are genuinely and legitimately concerned with both aluminium and fluoride.

Aluminium in so far as it is a problem, is an environmental one. So far no one has advocated adding the stuff to the water for dental purposes, though aluminium is used in medicine. But, who knows? One brilliant dental surgeon in Britain is using an alumina compound for making porcelain crowns. They are exceptionally strong and indistinguishable from real teeth. This dental worker is a modest man, though a remarkably excellent surgeon, and not given to excesses, but perhaps one day when nobody has any teeth at all, someone else may find a way of growing aluminium teeth in the cradle if only all babies can get it in their water?

Such amusing fantasy may sound flippant, but then, to anyone who has studied the medical literature on fluoridation, fluoridation sounds just as mad. Newcomers to the game should know that from time to time aluminium will be dragged into it, and the language used by both sides will sound ominously familiar. It will help him to cope more easily when the day comes when strontium, lithium, and fertility control agents are also added to the drinking water.

Returning to the more immediate problem of fluoride, there has been another and purely technical problem. In Britain, water engineers have been strangely quiet except for the odd rebellion such as the one in Edinburgh. But even such small British rebellions are not accompanied by grand public statements, resignation from professional organizations, and a lifetime's devotion to putting the message across to the public in the American manner. In Britain even water engineers are gentlemen. They keep quiet and hope that if they do nothing – except adding the stuff to the water – the problem will go away. In America they do things differently.

Admittedly, in 1972 one water engineer from a large British city then recently fluoridated, told me sadly over the phone, 'We are not used to being given an unbiased view in this matter from the British medical profession'. But that is as far as it goes.

In America they have Dr Willard E. Edwards. As is usually the case with people such as Dr Edwards, he appeared at first not to exist, and it took me some months to trace him to Honolulu. The US National Association of Corrosion Engineers claimed that he – like Dr Rapaport – was a figment of my imagination, though they did not reply to my last letter when I informed them that he was alive and well in Hawaii and writing nasty letters about them.

'Corrosion due to fluoridation is something the N.A.C.E. will *not* admit,' he writes and adds, 'I think the N.A.C.E. know very well who I am. I severed my membership with them (and they know why) when I found they are largely sustained by chemical and aluminium companies which sell or have fluoride as a by-product of their industries'.

Yet, even corrosion of water equipment appears to the faithful to be something which does not exist. They acknowledge certain practical difficulties, and even explain the detailed safety precautions needed to protect the men who handle the stuff at this stage, but if the antis complain, the argument is turned on its head: only a scientific simpleton would get mixed up about large amounts and small amounts. Lots of chemicals are dangerous in large amounts, and harmless or necessary in very minute amounts. Why, even rhubarb is dangerous if you eat enough of it. Certainly rhubarb, if it had been invented today, would not have passed the stringent modern drug regulations. Why, even ordinary salt could be dangerous if you took too much. Of course, people don't drink salt water, but that has nothing to do with it.

This argument can be made to go round in circles like a perpetual motion machine, and those who try really hard and sincerely to justify fluoridation on scientific grounds appear forever in danger of biting their own tail. They have had this problem from the beginning, and it was therefore necessary right from the start to concentrate part of the fluoridation campaign on converting water engineers, and one or two of them loom large in the early publicity. Despite this there has been perpetual trouble from this quarter, especially in New York where one leading opponent among the engineers is said to have exclaimed that fluoridation would be introduced over his dead body – which was almost what happened. The most solid opposition has come from Germany where the water engineers have made it clear that they would under no circumstances agree to fluoridate. On the other hand, this particular part of the fighting has gone relatively unnoticed by the public. Water engineers, after all, know nothing about dentistry, and their objections could be regarded as a minor and purely technical problem.

How minor? It is difficult to judge the situation in Britain. Those with the necessary technical knowledge will not be quoted or will not speak up at all. The whole fluoridation discussion in Britain has been bogged down with a traditional fear of being the odd one out. One doctor wrote to me about the 'minor technical difficulties' in Shropshire:

'Owing to the distribution system (of the water supply) it would only be possible to fluoridate by introducing fluorosilic acid into the system at 22 different points. The acid would be conveyed to these points in 200-gallon tankers, stored underground in 600-gallon containers, and stored on the surface for day to day use in buildings which would be unguarded from either vandals or children. Two fail-safe devices were to be fitted at each of the 22 stations, and another fail-safe device at each station to transmit a radio signal to a central control point to alert staff of any malfunction.

'From my knowledge of the very dangerous properties of this acid, as I had been a surgeon under the Factories Acts, I considered the whole scheme extremely dangerous, especially as these 200-gallon tankers would travel down the narrow lanes of Shropshire'. He added, 'I have now spent the last nine months checking and reading papers (on fluoride) by world authorities of great repute, and I am convinced that one of the greatest hoaxes in the history of medicine is at the moment being perpetrated'. He also complains that he had to write off to the *Swedish* Minister of Health to obtain the full facts because the Department of Health here only supplied memoranda 'with all the important and relevant facts omitted'.

From time to time anti-fluoridation dentists have complained to me that their water engineer friends are jolly tired of having to rush round and check the fluoride levels and that these cannot be properly controlled. But that is mainly hearsay. At present there seems to be a blanket on official information of this sort in Britain.

Our Honolulu man, Willard Edwards – who claims that he has worked for twenty years in the field of corrosion control and prevention, seven of these in fluoridated areas, and that he was employed from 1953 to 1965 by the US Navy as the corrosion control engineer for the Pacific area, as well as doing similar work for major oil companies in Hawaii – has strong views on the subject.

Fluoride, he says, acts differently in the water of different cities because the chemical content is seldom the same. In general fluoride has a great affinity for iron oxide. It often softens previously hard pipe scale, or the iron oxide surfaces of pipes and tanks. The softened scale is loosened and carried away from its previously-fixed location, thus allowing new iron oxide (rust) to form and corrode and weaken pipes and tanks under pressure. He claims that he has seen large increases in corrosion of pipes and tanks after the water had been fluoridated.

He also claims that the reason so many communities in the US who have tried fluoridation and given it up, have done so for exactly this reason. Corrosion was also the problem quarrelled over in the New York battle in the mid-Sixties, though the New York engineers had also concerned themselves with other scientific arguments.

Dr Arthur Ford, former Water Commissioner of the City of New York told the New York Board of Estimate that:

'The water supply system of the City of New York is not a gigantic medicine bottle into which one may combine ingredients and shake well before using. We control the concentration (of fluoride) going into the water at the beginning. No one knows what concentration will reach all the householders except that it will be different all over the City.'

'What benefits may result in reduction of tooth decay have been grossly exaggerated by erroneous and promotional handling of the fluoridation statistics in a way that would never be tolerated in an engineering office.'

The late Benjamin Nesin, Director of Laboratories, New York City Department of Water Supply, and an authority on water-supply toxicology, said:

'Never in the history of water supply has a substance with so much unfavourable evidence been considered seriously for introduction into potable water.'

Since then New York City is said to have had constant trouble with its mains and the water there has occasionally been reduced to a trickle. In Riverhead, Long Island, people reinstated and gave up fluoridation three times before abandoning it for good because of pipe damage. In Idaho even the water company came out against fluoridation:

'Excessive corrosion appeared throughout the system within a few months after the introduction of fluorine was started . . . fluoridation of the water was a very expensive undertaking for the Idaho Water Company . . .'

San Francisco and Schenectady are said to have had similar problems. Concord in New Hampshire had the same experience, but gave it up for good the first time round, footing a bill of 200,000 dollars for a new water system. In Wilmington, Massachusetts, the superintendent of the water system notified the townspeople in 1962 that since the installation of the fluoridator there had been a series of breakdowns of the equipment due to corrosion as well as an increase in corrosion throughout the town. It had also been impossible to maintain the required level of 1ppm, with the actual level often being 1.5 ppm. In South Bend, Washington, the Mayor stated in 1967 that the council had forbidden fluoridation. The mayor himself had originally campaigned in favour of fluoridation having been led to believe that it was calcium fluoride that was going to be added. That was not what happened. This mayor, Ivan W. Ginther, obviously feeling that he had been misled, later wrote to an anti-fluoridation group that he regarded fluoridation as 'one of the biggest and most dangerous hoaxes ever perpetrated on a people'. In Australia there have been similar reports of corrosion, but also reports that the official health departments have actually refused to examine corroded pipes. Everywhere the word 'hoax' is heard, and the sheer weight and volume of this endless stream of 'testimonials'—even if only half of them were true—would merit a major worldwide political investigation into the way this issue has been handled. Seldom have so many people in so many different parts of the world been in such remarkable agreement about what has been done to them.

In the US, the Environmental Protection Agency is

quoted as reporting that it has found inadequate equipment, maintenance, operator training and surveillance in fluoridation programmes in all the states surveyed and that in only 43 percent of the water systems were fluoride levels within the limits recommended by the state health or environmental authorities.

Even the *Journal of the American Water Works Association* has had to admit that 'silicofluoride solutions have been found to destroy steel pipe very rapidly' and points out that hydrofluosilic acid should only be handled with rubber or plastic materials. Yet the journal adds, 'the manufacturer's problem is disposing of (this) acid that has been produced. He must ship it, sell it, get rid of it'.

That was in 1951. In 1965 *Chemical Week* wrote, 'competition is so keen the middleman has been squeezed out, and producers deal directly with city purchasers'. The following year *Dunlop Dimensions* wrote with pride that:

'The Electric Reduction Company of Canada Ltd has constructed a plant which will manufacture more than 60 tons of hydrofluosilic acid every day. Already the plant is supplying fluorides for the water supplies of Toronto, Oakville, Belleville, Welland and Picton, and exporting to such cities as Syracuse, Rochester, Pittsburg and Chicago in the US . . .' In the past, hydrofluosilic acid had been only an incidental by-product of regular phosphoric acid production at Port Maitland. It was neutralized and simply thrown away. Now it was retained in special rubber-lined storage tanks and was, the reader was told, so corrosive that without the protection afforded by the rubber linings, the steel tank structures would be eaten away in a matter of hours. The future for sales of hydrofluosilic acid looked extremely good and the Electric Reduction Company of Canada had seen a market facing them and built a plant to tap it.

Another year on, in 1967, the *Oil, Pain and Drug Reporter* enthused about the fact that the ten big cities in the US using the acid, consume more than 7,000 tons annually on a 100 percent basis, and that there would be an almost unlimited supply of acid to meet all future fluoridation needs. The article also talked hopefully about state laws making fluoridation mandatory everywhere.

In the absence of reliable information from the authorities on fluoride pollution, one cannot perhaps blame the antis—who have few resources for doing their own scientific tests—for also making extravagant claims in this field. It is frankly impossible to judge many of these claims, but some deserve at least to be mentioned.

Thus American anti-fluoridation literature tells us that hydrogen fluoride was blamed for a killer smog in Birmingham, Alabama, in April 1971, partly perhaps because the authorities had claimed that they had no ozone or sulphur dioxide problems in that city. The smog was blamed on local steel, iron, and cement plants, especially the steel mills, and the mayor tried in vain to shut them down while it lasted. Eight deaths were recorded as a direct result of the smog. In San Francisco, on the other hand, the authorities are said to have shut down the Pacific Steel Co for 'trespass of fluorides'. In Oregon, a local aluminium company is reported to be wiping out much of the vegetable and fruit crops, and in Duluth, Minnesota, a steel company is reported to have been sued for 10 billion dollars over fluoride corrosion of a bridge. In Los Angeles the University of California Air Pollution Research Center is claimed to have admitted that 'it has been known for decades that hydrogen fluorides and sulphur dioxide are the principal culprits'. Hydrofluoric acid is also used in crude oil refining and in the processing of motor fuels, and this again is said to add to the smog content. How much is retained in the fuel and how much gets released as motor exhausts has never been revealed.

Marier, in his report *Environmental Fluoride* states that 'in several surveys in which sulphur dioxide had been suspected as the primary air pollutant, fluoride was found to be the factor responsible for environmental blight'. He points out that industries that release fluoric effluents also use fossil fuel as an energy source, thereby emitting significant quantities of sulphur dioxide, and comments on possible synergistic effects. 'Synergistic' means that a substance stimulates and enhances the effect of another substance. Thus, if the two occur together, the combined effect would be greater than the sum of either occurring alone. It is a phenomenon well known in pharmacology, but it does not appear to have been seriously considered in connection with fluoride from the medical point of view. So far only environmentalists have looked at it.

In 1966, Exner, in a review of the history of fluoridation, stated that when the policy of fluoridation was introduced, the pollution situation was 'desperate'. Thus, he claims, Peace River in Florida from which Arcadia takes its drinking water, had up to 17 ppm of fluoride caused by superphosphate plants in the river basin. He also claimed that a lawyer for a leading copper company had stated that Salt Lake City would be fluoridated whether the people liked it or not—'how else can we get rid of our fluorides?'

Exner also tried to link the London killer smogs in the 1950s with fluoride, particularly with fluoride-containing motor fuels, though not everyone would agree with him there. He compared London with areas in Washington which are equally noted for their natural fogs and which are heavy with sulphur pollution but have almost no fluoride pollution and never experienced real smogs. Researchers at the Washington State University seemed to be thinking along similar lines at one point and conducted studies in Washington, Oregon, and Utah—and apparently came to conclusions which were different from official studies or those conducted by committee fluoridation-promoting organizations. In 1961, the grants from the National Institutes of Health to that university were drastically cut, while grants to the second university in the state, the University of Washington, which had conducted no research into fluoride since 1951, were even more drastically *increased.*

In an article in the January 1966 issue of *Aqua Pura,* an Australian anti-fluoridation newspaper, Exner stated that 'it is now clear that the one utterly relentless force behind fluoridation is American big industry, and that the motive is not profit as such, but fear'.

Fear of what?

The first recorded lawsuit for damages against a fluoride polluter was in Germany in 1855. The Freiburg smelters had paid out 880,000 DM by 1893 for current injuries and 644,000 DM for permanent relief. Fluoride-induced disease of cattle in the area was endemic. Around the turn of the century the fluoride pollution was becoming acute in both Germany and Britain, according to anti-fluoridation researchers.

The full chemistry of the problem does not seem to have been understood until the second or third decade of the present century. In 1933, Dr Lloyd DeEds, senior toxicologist with the Department of Agriculture and a lecturer on Pharmacology at Stanford University, warned about the serious nature of fluorine toxicity and especially about the long-term effects. He even warned against baking powders, 'if carelessly manufactured', and claimed that they might then contain as much as 0.5 per cent of fluorine which would provide a daily dose of between 4 and 35 mg.

By 1939 it seemed that industry had become acutely worried about the situation and the problem was put to the Mellon Institute. Just then Gerald Cox came up with his famous sentence that '. . . the present trend towards complete removal of fluoride from water and food may need some reversal'.

It is this 'reversal' which we are living with, and fighting over, today. Anti-fluoridation literature reverberates with accounts of lawsuits for damage to animals and crops, but it seems that there has been only one successful suit in America for damage to humans: the Paul Martin Family case against the Reynolds Metals Company, though the company in question obtained their medical testimony from the Kettering Laboratory, and a whole string of other major industrial concerns joined them in support.

The really serious increase in fluoride pollution has occurred since the Second World War, and greater pollution must be anticipated with the spread of nuclear production. Both hydrogen fluoride and elemental fluorine is used for the production of uranium. Exner points out that while pure elemental fluorine was made only with great difficulty prior to 1942, in minute quantities and 'could not be bought at any price', it is now shipped around in tank-trucks of 5,000 lb. capacity.

A National Academy of Sciences report in 1971 states that cryogenic trailers have been developed for shipment of bulk amounts of liquid fluorine by truck. One type of trailer now in use has a capacity of 2.5 tons of liquid fluorine. The fluorine is contained in an inner vessel surrounded by liquid nitrogen and an outer insulating jacket. 'Because fluorine is one of the most reactive elements known and is highly toxic, great care is required in loading, transporting, and unloading the liquid cargoes.'

The possibility of accidents, as always, tend to worry laymen and the uninitiated more than the experts. Exner speculates on what would happen in case there was an accident: there would be no explosion, but the fluorine gas would consume everything it touched, including water, steel, concrete, and people. The heat would be terrific. The products of combustion would all be poisonous, and most of them corrosive. Enough poison to kill a million people would result, and decontamination would be a major undertaking.

I tried to check this with British Nuclear Fuels Ltd in Risley, Warrington, who provide uranium to the Atomic Energy Authority. The conversation went like this:

'Disposal? Disposal is simple . . . there isn't a disposal problem in the real sense of the word . . . these waste products are becoming increasingly valuable . . . we are looking at ways we can utilize them . . . one ends up with either a solid or a liquid waste, and in the case of liquid waste one pumps it out in the river . . .

'Transport of elemental fluorine gas? We don't transport it, well, not in any great quantities, in fact we don't transport it, we generate it on the site. . . . No, none is transported in Britain for the nuclear industry, though of course others might do so for other industries. Which industries? I really don't know of any. I think it is used for rocket fuel, but I'm not sure . . . I'm frightfully busy with something much more important for the BBC, would you like to write me a letter so that I can answer your queries in detail, I should also like to consult the management about these . . .'

A short while later, the BBC Science Features Department produced in their Controversy series a programme in which the Chief Executive of this firm argued that nuclear power production was more or less absolutely safe. No one breathed a word about fluoride waste problems, nor about transport of elemental fluorine gas. The problem, it seems, does not exist. No figures are published for the production of hydrogen fluoride or elemental fluorine, and it would be fair to say that the public at large is totally ignorant of the many different industrial processes that use fluoride in one form or another.

The National Academy of Sciences, in its 1971 report on fluoride air pollution, listed areas in which knowledge of the possible toxic effects of fluorides on humans is inadequate. It stated that 'there is a paucity of data in the literature describing the effects of fluoride on the human heart and cardiovascular system' and 'there are very few data available related to the effect of fluoride on the central nervous system'. In 1968 Professor H. C. Hodge, from the School of Medicine and Dentistry at the University of Rochester said that 'no specific large-scale epidemiological studies are available comparing the health of fluoridated communities with that of communities where the water supply contains only traces of fluoride'. Evidence of what effects the continuing increase in general fluoride pollution from many different sources might have in future, is equally hard to come by. Yet in 1972, the US Public Health Service declared that 'special health studies beyond those that have already been completed are not contemplated'.

In 1972 Michael J. Prival, Ph.D. and Farley Fisher, Ph.D. wrote two reports—*Fluorides In the Air* and *Fluorides and Human Health*—for the Center for Science in the Public Interest in Washington. They should be required reading for anyone who sincerely believes that 'there are absolutely no ill effects' and that 'the crank oppositon has now died down' as Bruce L. Douglas, Professor of Public Health at the University of Illinois, wrote in a letter to the *Sunday Times* after my article in that paper on fluoridation in September 1973.

On the contrary, it seems that opposition among the scientific community in the US is growing rapidly. While the effects on human health are still very debatable, there is no denying the fact that agricultural damage is exceedingly well-documented, and it would not be possible to continue to ignore this documentation forever. Strong words come from even the most respectable sources. Thus Dr Howard H. Hillemann, Professor of Zoology at Oregon State University put on paper the following passage which is not the sort of thing you often get from British science writers:

'The power-pack of posturing pseudo-scientists and medico-dental politicians, well greased and supported by federal commissars, act reflexly in the spirit of Orwell's *1984* to use our tax dollars in brazen attempts to impose by means of their unconstitutional and perverted moral concepts, mass medication via poisonous fluoride in Oregon's domestic waters—all without a sliver of scientific evidence.

'We stand adamantly against this outrageous nutritional faddism and shamanistic medico-dental quackery, as fostered in certain bastions of political power defending the citadel of systemized ignorance.'

'Egads, what a subject!' says Ralph Nader, 'You're either for it, or you're against it. If you're for it, you are a Public Health Service hero. If you're against it, you are considered a kook.'

'Kooks', 'frauds', 'hoaxes', 'power-packs' – and the establishment claims that there is no opposition.

Michael Prival adds a nice human touch to a story from Montana. When Ed Mollenberg sold 54 acres of his land in Garrison, Montana, to the Rocky Mountain Phosphate Company, it was a matter of civic pride. The new factory promised jobs and tax revenues for the industry-hungry region.

Four years later, the pine and fir trees were turning brown and the cattle were so crippled that they could not stand up. Ed Mollenberg stood up before the citizens of Garrison and cried, 'May God—and the people of Montana—forgive me for the mistake I made in selling my land to such an industry!'

In January 1970, after six years of vigorous and frustrating campaigning to save their town, the residents of Garrison succeeded in forcing the Rocky Mountain Phosphate Company to close its doors for good and were awarded a 123,000 dollar judgement against the company.

'The story of Garrison is but one of many in which an industry and its neighbours have fought long and bitter battles, both in and out of court, over damage caused by fluorides. The federal government and many affected states have failed to enact any regulations to control fluoride air pollution. They have taken action only when public pressure has forced them to recognize an emergency. The people of Garrison had to fight hard to shut down an industry that threatened the very existence of their town.'

Dr Clancy Gordon of the University of Montana commented, 'I believe the reason the bill (to set air pollution standards in Montana) actually passed was the dedication of Garrison's residents and the pictures of the crippled cattle on their knees'.

Garrison has now become part of American fluoride folklore, and the sight of crippled cows crawling across pastures on their knees is indeed heart-rending. About the effects on people we know less, and perhaps more seriously in the long run, we do not really know how scientific or political crazes such as this one catch on in the first place.

The real trouble with the *fluoridation* issue and the dental side of things lies not so much in the scientific merits or drawbacks. Fluoridation has not spread because of its merits—a difference of 0.9 of a tooth would not impress many people. Nor has it been always opposed because of its dangers since the public knows little of those dangers.

The remarkable success of the fluoride promoters, not in preventing tooth decay, but in attracting new converts to the fluoride creed is not unlike the spread of great religions or political beliefs.

Like these the fluoride faith thrives on its own contradictions. Contradictions add an element of mystery and poetry, and mystery requires faith and fables, not scientific analysis. Fluoride is dangerous, yes, but it is also safe, safe and pure and wholesome as milk and your teeth will be like mother-of-pearl.

Faith must be constantly nurtured and sustained by the emotions, it must be proclaimed and exhorted with long and difficult words, there must be a secret language. It is little use addressing such emotions and such fables with facts and figures and statistical graphs.

If this is the way scientific controversies will be 're-solved' in future, it is an ill omen. Perhaps it is no coincidence that this kind of modern scientific faith should emerge at a time when traditional religions are losing their grip, and when even the worshippers of Eastern types of mysticism find it necessary to resort to the terminology of nuclear physics to explain their ideas.

While the few may find solace in philosophical relativity, the many would require something simpler. And fluoride is nice and down-to-earth. You encounter it every morning in your bathroom in your toothpaste, though you may not know that you have it too in your frying pan and your hairspray.

In the end it is a remarkable ploy: give fluoride to the people lest they discover that they already have it.

VII

Acid Arguments: Alternatives

> "It is truly remarkable how the same old people can say the same old things so many times in so many different ways . . . their obsession with this theory is one of the most incredible love affairs in the history of nonsense . . ."

Fluoride is not the only controversy in dentistry. One can hardly scratch the surface of the official dental party line in the English-speaking world without finding the place teeming with wailing outcasts, refugees, and would-be or failed revolutionaries. Not all of them are cranks. Some hold or have held the most respectable positions. Some are practising dentists. Others come from different fields of science such as chemistry or soil microbiology. It may be no coincidence that the best alternative ideas to present thought in the realm of prevention have come from soil microbiology, since the whole of the alimentary canal is swarming with bacteria which we could hardly do without, and there are some heretics who do not understand why the bugs should be eliminated in the mouth if they must be preserved elsewhere.

The opposing theories vary, but many have one thing in common: the familiar claims of obstruction from normal outlets in the usual and appropriate journals. As so often with medical heretics, many become extreme and unreasonable in their views, retiring to their own self-selected wilderness or joining that extraordinary collection of men, the modern medical 'underworld', whose theories without exception are all 'absolute nonsense' and who have all without exception been totally 'discredited'—if orthodoxy is to be believed.

There is nothing unique about dentistry in this respect, nor even about fluoride, though dentistry has gone further than the rest of medicine in the suppression of apparently unacceptable ideas. Fluoride is only one of many dental controversies and, from a purely dental point of view, some of the others are far more interesting. It is possible that if some of them had not been so totally 'discredited', the present generation of children might have had far better teeth. At the same time fluoride cannot be separated from these other controversies.

The quote at the start of this chapter comes from Schatz, as some readers might by now have guessed. Schatz is the author of one of the alternative theories of dental caries which orthodoxy has not been able to ignore completely, in so far as it is mentioned in all good dental textbooks and a somewhat subversive interest in it continues to this day.

The 'incredible love affair' he talks about is the so-called 'Acid Theory' of dental caries, a theory which emerged during the nineteenth century, was firmly put on the map by W. B. Miller in 1890, and has hardly changed since. Today no dental student would dare challenge it, and most adult working dentists who may have private doubts about it, keep their doubts private. There can rarely have been a medical theory which has been accepted in such a totalitarian manner, practically as an act of faith, and yet survived for so long. It says something for the discipline, if that is the word, of orthodox dentistry that dissenters should have been kept down for so long and so successfully. Leading dentists like to take pride in the almost global co-operation which pervades modern dentistry – and their unhappy opponents see in this global system something amounting to world-wide 'dictatorship'.

'The main reason,' says Schatz, 'why there has been so little real progress is the continuous effort to avoid controversy. Real free and open discussion has been avoided. At times, such discussion has been vigorously suppressed. This has become clearly obvious in the United States where symposia, conferences and congresses on dental caries have literally become epidemic in recent years. Unfortunately the reputations of most dental authorities have been built on the acid theory. Since their stature in dental research now rests on this theory, they have carefully excluded in their selection of invited speakers those who would and could seriously challenge this theory. Consequently, at most such meetings, the time-honoured members of a mutual admiration society repeatedly compliment one another for the little work on acid they did many years ago and have been talking about ever since . . . during this century nothing of any consequence has been accomplished.'

Schatz is right in so far as *prevention* is concerned. On the surgical side of dentistry one can have nothing but admiration for what has been achieved. Almost anyone who can afford it, can have his mouth restored to the point where only the expert could tell the difference, and in this respect modern dentistry combines the best of technology with what can really only be described as good old-fashioned craftsmanship.

But the sad truth is that after a century of otherwise brilliant scientific advance, virtually nothing has been achieved in the field of prevention. People's teeth are getting steadily worse, and dentistry is dominated by commercial interests to such a degree that it can be difficult to distinguish science from plain advertising.

The reason is that the claims of the dental oral hygiene market are closely linked with the Acid Theory. It is a profitable theory, and it has, in an odd way, helped promote the current expedient belief among politicians that dental science is a kind of extravagant cosmetics and not a part of 'real' medicine. How awkward this theory can be for the experts and how vulnerable it is to humorous quips from students, has been marvellously illustrated in a quote from the *London Hospital Gazette*:

'A Registrar at the Urology Department was heard to put forward the theory that dental decay is caused by little germs with pick-axes. Perhaps he also belongs to the school which thinks that urinary calculi are caused by little germs with cement-mixers, in which case one can see the Urology and Dental Departments getting together, so that dental germs with their pick-axes get to work on patient's urinary calculi, while the Dental Department gets those little cement-mixers working on filling their cavities.'

The Acid Theory states that caries is caused by *bacteria in acid fermentation.* This was an impressive idea at the time of Pasteur, and there is, of course, not the slightest doubt that once you have a cavity or 'dental wound', there are plenty of bacteria around producing acid. Nor does anyone doubt that the disease is linked with fermenting carbohydrates.

But that does not prove that this is how decay starts in the first place, especially as decay starts in the dentine below the enamel. Attacks on the teeth from without have been thoroughly researched, but resistance to attack from within is poorly understood.

Dental science is an extraordinarily complicated and exciting subject, and caries is more complicated than lay people would at first expect. Moreover, because teeth take so long to erupt, they are extremely difficult to research experimentally, and there are good reasons why dentists should have been on the defensive. The teeth are the gateway to the rest of the body, and to some extent mirror the state of the rest of the metabolism. They cannot be treated in isolation, and in some ways they are almost too much for dentists to cope with alone. Yet even doctors are often quite remarkably ignorant of the physiology of the mouth, and if dentists had to cling to any one aspect of dental caries in order to prove that they were at least doing something, acid was perhaps the most 'scientific sounding' candidate along a lot of others.

All the same the great fear of acid is a bit odd. For a

start an ordinary apple can be up to fifty times more acid than any acidity found on the teeth, and even a tomato is more acid. So deeply ingrained is the fear of acid that I recollect hearing one dentist warning people not to eat apples. The fact that teeth do not crumble instantly when you bite into an apple did not appear to make any impression on his belief that acid is bad. A lot of healthy food such as fruit is, in fact, extremely acid.

Jenkins has stated that caries can occur in the absence of acid, and that acid production in itself would not necessarily cause dental decalcification. It has also been pointed out that although bacteria are always present in the mouth, injuries usually heal rapidly and rarely become infected. W. H. Bowen, from the Department of Dental Science at the Royal College of Surgeons, mentioned in the *British Dental Journal* in 1971 that there has been a failure to implant human strains of micro-organisms in the mouth of experimental animals. Yet research is usually done on such animals. So far between 25 and 30 different strains have been identified – an awful lot to juggle around with for those who hope to develop a dental vaccine for humans.

Coupled with the fear of acid and bacteria is a closely-linked fear of dental *plaque* – a word which has already become commonplace in toothpaste advertisements and which no doubt helps to make the ads sound more 'scientific'.

Bacteria, we are told, breed in the plaque or slimy layer which covers the teeth. The logical conclusion is to try and get rid of the plaque. But it is known that dental plaque is found everywhere in nature, and some researchers claim that dental plaque may even serve some purpose in protecting the teeth against the *alkalinity* of the mouth and the saliva.

It does seem probable that the plaque found in the mouth of an animal or person who lives on raw and fresh food, is very different from the one found in a person who eats fudge and sweets, and that the real culprit is not the plaque as such but its constituents. But this would once more put dentists in one of their awkward dilemmas. If the sweet-and-soft food type of plaque should be removed to avoid damaging the teeth, then the teeth would also be left without the protection of a more 'natural' plaque.

It suits the toothpaste manufacturers that we should believe that plaque ought to be removed. The word *dental plaque* has now become part of the great dental myth and is carefully indoctrinated into the minds of school children, with endless ceremonies of coloured disclosing tablets to show them what filthy mouths they have, and with reading matter explaining how valiant dentists devotedly wage a splendid war against this Enemy Number One. The disclosing tablets may do some good – our children do have filthy mouths. But plaque remains an awkward problem.

There are other flaws in the Acid Theory. A Danish dental researcher, H. Eggers Lura, claims to have shown that an acid reaction in the mouth is harmless and so is fermentation. *Concentrated* sugars, he points out, form little or no acid. Sugar does the most damage in the highest concentrations, yet the presence of bacteria in such high concentrations is almost nil. If you believe in the Acid Theory, you therefore end up with another impossible contradiction: a piece of sticky and highly concentrated sugar may lie up against the tooth where the damage is obviously being done, while the bacteria flourish happily in the diluted saliva solutions furthest away from that tooth.

Eggars Lura claims that sugar may have a direct chemical reaction with the surface of the teeth, without the bacteria. He also thinks that researchers have failed to distinguish sufficiently between the initial attack and the well-developed dental 'wound', and that too much has been deduced from what happens in the saliva, though both the acidity and the bacterial flora in saliva may be different from that found in the plaque. Dental science, in other words, has been seriously hampered by the fact that you cannot lock up a living mouth in the laboratory, and much of the evidence on which the present theory is based is deduced from animals, from dead teeth, from saliva – though the attack is on the living teeth in the plaque.

More seriously, the Acid Theory has been challenged by Schatz, and before him by an American dentist called Charles F. Bodecker.

Back in 1905 and 1907 Bodecker established the presence of *organic* matter in the enamel. The Acid workers only paid attention to inorganic mineral matter in the tooth structure.

Bodecker's discovery paved the way for the rival theory that caries sets in, not in the mineral components, but in the protein matrix, and that organic matter can degenerate beneath the surface of the enamel before caries and demineralization occurs. In the 1950s there was growing interest in this approach, but as the action of fluoride could only be explained at the time with the Acid Theory, the two have soldiered on hand in hand ever since. Schatz claims that Bodecker himself had an article on the subject rejected by the *Journal of the American Dental Association* only weeks before he died.

In 1953, Schatz and his co-workers came out with a further development on Bodecker's protein idea and produced the so-called Proteolysis-Chelation Theory of caries. It was published in the *Journal of the American Dental Association in 1965*, though an earlier version had been published in Spain in 1954 after – according to Schatz – the Americans had originally refused to carry it. Schatz also claims that while he had received funds from the US Public Health Service and the New York Academy of Dentistry for the work, he was told to stop work on protein or the funds would be withdrawn when his opposition to acid was discovered. The grants did stop, and he is now accused by his opponents of being only an 'armchair researcher' in this field.

The theory, however, survived, and was presented yet again in 1972, this time in the *New York State Dental Journal* with an editorial note recommending it as 'a concise explanation of this too-neglected theory by its brilliant originator'. Back in 1963, at a symposium devoted to the theory, it was stated that 'it seems that we have reached a stage where Miller's (acid) theory and the proteolysis-chelation theory are now diametrically opposed to one another. Only future work can decide which of these two theories will provide the long-sought explanation of the decay process responsible for caries'.

Schatz and his friends do not see it in quite such antagonistic terms. They feel their theory would explain *all* decay reactions under acid, neutral, and alkaline conditions in the mouth.

Proteolysis means the decomposition or breakdown of proteins. Chelation comes from a Greek word meaning 'claw' and describes a process where unwanted metal ions may be pinched and 'locked up' in other compounds. Chelation is widely operative throughout nature in soils and in living systems. Thus haemoglobin in the blood is an iron chelate, and chlorophyll is a magnesium chelate. Enzymes which require trace metals such as zinc, copper, iron, cobalt, and manganese operate as chelate systems. Chelation, according to Schatz, is also the means of decalcification in the caries process.

With his background in soil microbiology he charges that many dental researchers simply don't know enough about chemistry to understand the complicated chelation theory.

'We were perplexed,' he says, 'in 1953 by the seemingly endless number of publications in the pH (acid measurements) of saliva, dental plaques, and fermenting debris in cavities. Tooth decay is not a disease of saliva, dental plaques, or fermenting debris in cavities. It is a disease of *teeth.*'

Yet today, 'it is still acid, acid, acid, acid, with the great majority of dental researchers who continue to use their pH meters more than their heads.'

Typical language from Schatz. He even goes as far as claiming that Miller's disciples have converted the original Chemico-Parasitic theory from 1890 into a virtual Acid

Theory, thereby restricting and distorting the old master's views.

'It became so easy and fashionable to measure acid that dental researchers became infatuated with the technique. They made endess pH measurements and rediscovered that fermenting carbohydrates give rise to acids. With such experiments repeated *ad infinitum*, they convinced themselves that acid causes caries. The pH concept may therefore be more responsible than anything else for the lack of progress in our understanding of this disease and our efforts to control it. And for the fact that a century of caries research had been doomed to failure from the very start . . .'

The proteolysis part of the theory claims that the *earliest* change in tooth decay is a rupture or breaking of bonds linking the organic and mineral components of enamel. The combined theory states that the whole complicated business is caused by an enzymatic attack on organic constituents and a more or less simultaneous demineralization initiated by substances able to complex or chelate calcium, and that this arises from the degradation of the organic constituents. Hardly the stuff toothpaste advertisements are made of.

There are a number of objections to this theory. On the other hand, recent work at the University of Alabama Institute of Dental Research has shown that protein deficiency during gestation and lactation may significantly increase susceptibility to tooth decay – though the researchers in question seem to think that this has something to do with proteins in saliva and appear unaware of the presence of protein in enamel. At least they do not mention it. Another group of researchers, at Tufts University School of Dental Medicine, have come up with a new compound for removing tooth decay without drilling by breaking down dead protein in the enamel. It is all a very complicated business, but the day may yet come when at least part of Schatz's theory may lend itself to the copywriting treatment.

From a very broad point of view, protein is important

because it brings everybody right back to where they started, to the fact that there is no escape from nutrition as a key factor in healthy teeth.

There is in the whole nutritional-dental field a lot of sadly neglected research and incredible confusion. Dentists are not trained to be nutritionists – and it shows.

How confusing is illustrated by work at the Case Western Reserve University in America which suggests that *chewing* is bad for the teeth because, this theory says, caries is caused by enzymes released by white blood cells migrating from the gums. The harder you chew, the more you stimulate migration, and the cure lies in incapacitating the white blood cells. This should not be difficult – because research at the Science Resources Foundation, Watertown, Massachusetts, has shown that the gases from a single cigarette will kill off some 50 percent of the white blood cells in the mouth. Remarkably, no one has yet come out and recommended smoking as a cure for tooth decay.

Most workers obviously believe that chewing is good for the teeth. It would be odd if it weren't. Jenkins has commented on the seriously underdeveloped masticatory apparatus of civilized man. Chewing is probably the most neglected field of preventive dentistry today. There are modern children who can hardly chew their way through an ordinary carrot without aching jaws. And yet it appears that no one is even certain about what loads are put on the teeth during chewing. It could be anything from 20 to 200 lbs, and workers at Surrey University who have looked at this problem, have complained that dentists are simply not being scientific enough about it.

Chewing has a number of very beneficial effects. A tough, fibrous diet increases the volume of saliva and may have a direct mechanical action in preventing the accumulation of food debris and in removing bacteria. Caries has also been associated with lack of saliva (for instance in thyroid disease), and the removal of the salivary glands may increase the number of cavities in a startling way.

Saliva also contains an enzyme, lysozyme, which has anti-bacterial action, and white blood cells which, very usefully, have been found containing ingested bacteria in the mouth. That saliva is a useful substance has been vindicated by further work at the US National Institute for Dental Research which showed that darkness reduces the flow of saliva by as much as 75 percent below its normal daylight level – something which might help explain the devastation caused by the pernicious habit of going to bed with sweets in the mouth.

Not all liquids are helpful, though. Research on thermal stresses in human teeth at the University of Utah has shown that tooth enamel can suffer thermal fatigue when exposed to very hot or cold liquids. This may result in cracks in the enamel surface and increase the risk of cavities, especially during a sudden drop in temperature such as the one produced by chewing on ice cubes.

Basically, it is the chewing that matters. The gum tissue and the circulation are stimulated, and so is the underlying bone structure. Good, strong, broad jaws will produce a mouth with well-spaced teeth, something which in itself counteracts caries. Some researchers have also suggested that the regular chewing of tough food may help harden the enamel itself and influence the molecular structure of the teeth by compression. Possibly this was what happened to Palaeolithic man, who had a low rate of caries.

As late as 1933 a survey in Switzerland showed that people who ate stone-hard rye bread had only 0.3 percent cavities per person, while people in St Moritz, whose teeth were whiter and were cleaned more often, and who ate soft white bread, had 30 percent cavities per person.

There can be little doubt, generally speaking, that 'chewers' have better teeth than the rest of us. Or, put the other way round, that teeth like any other part of the body will tend to fall into decay if not used for the purpose for which they were intended.

The modern, super-sophisticated solution, put forward by at least one British dentist, that the answer to the under-used, over-crowded, decaying teeth of Western children, should be to pull some of them out, sounds almost ludicrous. That is what is happening to most of them already, and the result is not better-spaced teeth, but drifting teeth and shrinking gums.

Equal confusion is found in other fields. Even an apparently innocent subject such as cow's milk demands caution. The claims which have in the past been made by housewives and some dentists alike that withdrawing free school milk would result in wholesale dental destruction, are not exactly correct. Nor are the advertising claims that 'milk toughens growing teeth and helps them resist decay'.

While milk is an excellent food for older children, it has been known for decades that the ratio of calcium to phosphorus in cow's milk is wrong for the human *infant*. Though there is four times as much calcium in cow's milk as in mother's milk, less is absorbed. The real problem here is breast-feeding. Some 90 percent of British mothers still bottle-feed their babies, and from a dental point of view we are coping with virtually a whole bottle-fed generation. Yet research both here and in America has shown that breast-fed infants have better teeth than artificially-fed ones. Indeed, in the East in rural communities, children have grown up with splendid teeth without ever touching a drop of cow's milk. Yet the dental profession has made practically no attempts to inform the public of the dental importance of breast-feeding.

Nor have there been many official attempts to enlighten the public about the dental importance of early nutrition. In 1966 the Medical Officer of Health in Eastbourne did a pilot study which showed that a diet of wholemeal bread would produce only around two decayed teeth per child, as compared with the 1961 fluoridation figures which showed between six and seven decayed teeth per child in the five-to-seven age groups. Nobody seems to have taken much notice except a few individuals such as Mr Bernard Cooke, for forty years a Public Dental

Officer and one who complains bitterly that the British Dental Health Foundation simply ignore the questions and arguments he puts forward.

In Germany, at one point, the dental hierarchy went the whole hog and really showed their hand on this issue. According to Peter Bunyard, formerly a science writer with *World Medicine* and now one of the co-editors of the *Ecologist*, a Black Forest dentist, Dr Johan Georg Schnitzer, had organized a village of several thousand people and persuaded them to change their eating habits from typical consumer products to wholemeal food. The result, it was claimed, was that the incidence of dental decay among the children came cascading down.

Far from being given a medal, Dr Schnitzer was hauled before the German equivalent to the General Dental Council and threatened with being struck off the register. He was saved by the fact that a television crew happened to be in his room while he received an abusive telephone call from one of the dental pundits. This conversation was recorded and caused the German press to come out in favour of Dr Schnitzer. Nevertheless he appeared before the council's tribunal and was accused of practising his profession 'in the manner of a pedlar'.

There are a lot of dental pedlars around, all with their own pet subjects which they feel deserve attention. Strontium is one of these more specific agents.

Quite recently, Professor Itzhak Gedalia at the Hebrew University-Hadasah School of Dental Medicine in Israel, and Dr Fred Losee at the Eastman Dental Center in Rochester, US, have both been looking into the caries-prevention action of strontium occurring naturally in water (and not to be confused with the radioactive variety). Its importance, however, was pointed out already twenty years ago.

In 1953 a Norwegian dentist, Dr Hans Lodrup, and a chemist called Ottar Rygh, did a comparison of the occurrence of dental caries in Bonn and Oslo, quoting earlier German work. In the Bonn region more than 50 percent of children leaving school at the age of 16 were completely free of decay, while in Oslo almost all the children suffered from caries. The chief difference was found to be the mineral content of the water.

Bonn water, though low in fluoride, contained considerable trace amounts of strontium and vanadium. The Oslo water was poor in both chemicals. At the same time, it was found that the Oslo children with bad teeth drank more milk, while the Bonn children with good teeth ate more sweet dessert. Rygh came to the conclusion that 'the most marked lack of calcification is found when strontium and vanadium are absent at the same time'. The German scientists they worked with had come to the conclusion that there was no connection between the presence of fluoride in the water and caries. In fact the more recent research on strontium seems to indicate that good strontium-related teeth are found in areas which are *low* in fluoride. This, for instance, has been the case in Ohio.

There is talk already of adding strontium to the drinking water—but it comes from people who are also keen on fluoride. It is difficult to see how they combine the two theories.

This is just one example of neglected research, but it is clear that the fluoridation movement has had a devastating effect on most, if not all, such research ideas which happened not to fit in neatly with the fluoride philosophy. It is also an example of how much useful work done before and immediately after the last war has been neglected and often completely forgotten. Much of this work would obviously have to be subjected to modern scientific scrutiny, but this does not seem likely to happen while the most outstanding names in dentistry are totally and utterly committed to the fluoridation idea. Another unfortunate result has been the way in which the 'outcasts' in dental science, such as Lodrup, Schatz, and others, have formed almost a kind of 'alternative society' – being as much in touch with each other and supporting and quoting each other in much the same and sometimes biased way as the 'official' dental scientists.

It is very easy, once you discover one of these members of the alternative dental society, to track down all the others.

One much ridiculed dental writer – the late Dr Alfred Aslander from the Division of Agriculture at the Royal Institute of Technology in Stockholm – is now very highly regarded by some Norwegian school dentists who have actually put his theory to the test.

Aslander's writing was not all that scientific, but it made good common sense, and it seems that it works. He came up with the idea of Complete Tooth Nutrition, and advocated that people should eat bonemeal because a good quality bonemeal contains all the nutrients, minerals and trace elements thought necessary for good and strong teeth, including fluorine as well as proteins.

He had in his lifetime a long list of distinguished supporters including Professor Pierre Laplaud from the School of Dental Surgery in Paris, who wrote a thesis on the subject dedicated to Aslander. He provided, for a Frenchman, a remarkable quote:

'We have here a theory that explains all the known facts . . .' (though I would personally doubt that statement). 'It is in agreement with all the experimental data that we are able to obtain, and with data that have never been explained before. . . . It is the only way for providing good social dental conditions in a country. It will certainly be the most important improvement in human nutrition this century. It may prove to be of such medical importance and its prospective value is so great that I think it is the most important idea in medicine since Pasteur.'

British dentists and fluoride promoters called Aslander 'an old fool'. In one way they may have been justified in doing so, since his dental methods were far from being as scientific as his work in his own field, plant technology, may have been. But it is one of the problems in modern science that simple ideas often don't sound 'scientific' enough when put on paper.

Aslander claimed that bonemeal can prevent dental

caries – *totally* prevent it – if given to a child from birth and preferably also to the pregnant mother. He based the theory on personal observation and on historical evidence. Thus primitive man used to crush bone and mix it with his food, and in some parts of the Baltic region where formerly dental caries was unknown, the farmers ate the bonemeal they gave to their pigs. They also often ate fish whole, including the bones. Yet the mainstay of the diet of these people was barley bread and porridge, the sticky porridge being eaten with milk just before going to bed with no brushing or even rinsing of the mouth. They also rarely ate fresh fruit or vegetables. This was a poor man's diet, and it produced excellent teeth.

Aslander points out that the necessary minerals could of course be obtained elsewhere from other foods, but the mineral content of the soil and the water may make a big difference to such food. Poor teeth in North-Western Europe have been linked with the prevailing climate which tends to soak the soil and leach out easily soluble minerals, and similar connections between soil and dental caries have come from the US and New Zealand.

In Norway two school dentists, Hans Stoltenberg and Helge Myhre, have tried the bonemeal method and claim quite dramatic results – though the battle continues. Thus Myhre has an amusing little story to tell about attempts by the Norwegian health authorities to stop him giving bonemeal to pregnant women on the grounds that it contains fluoride, though these same authorities were campaigning in favour of fluoridation at the time.

Even this, however, is not the whole story. There are yet more quarrels, more controversies, more question-marks. One or two are perfectly 'respectable'. There is thus a conflict between the fields of biochemistry and genetics.

It is known that there is a similar distribution of caries, within the mouth, in different populations. Thus, surprisingly, the lower incisors often last longer than any of the other teeth though subjected to the same harmful influences as the rest of the mouth. Even on a single tooth there may be a genetic pattern of attack, and a particular type of attack on a tooth in the left side of the mouth may correspond to a similar attack on the same tooth in the opposite side of the mouth. In Britain, the leading researcher in this field is Professor D. Jackson at the University of Leeds. His research is not easy to reconcile with either the Acid Theory or conventional nutritional ideas. Nor for that matter with fluoridation – though the Professor is an outspoken supporter of fluoride.

There are yet other mysteries.

Dr Philip Sutton at Melbourne University has worked on the emotional side of dental caries for a good ten years, despite his involvement with the fluoridation issue. In a series of papers published in *Nature*, the *Journal of Dental Research*, and the *New York State Dental Journal*, he has described how an anxiety questionnaire can be used to predict fairly accurately whether the patients are likely to have acute dental caries. On one occasion he found that a high proportion of patients with this type of adult decay had a history of mental stress during the twelve months before they were examined. This was especially the case during a period 'which was a difficult one for businessmen due to an imposed restriction on credit'.

Any direct cause and effect relationship between cavities and stress and tension is not known and Sutton naturally considers that stress is only one of the many factors in the aetiology of caries. However there appears to be something special about the type of caries which develops under these conditions: acute or 'rapid' caries as opposed to chronic caries. This type is found in patients above the age of 25 and up to and above the age of 55 and sets in very soon after the onset of stress. Thus it could not have been caused by a modification in diet.

Acute caries is normally the sort which is encountered most frequently in children and in young adults and is characterized by a 'white halo' round the cavity while the colour of the affected dentine is yellowish white. In chronic or 'slow' caries normally observed in adults, the affected dentine is dark yellow, brown, or black. The intriguing thing here is that adults should display a symptom usually only found in the young. One feels tempted to ask what part stress might play in childhood caries. Usually caries is a disease of childhood and youth – the problem for adults is periodontal disease.

It is all a great mystery, but the observation is not new. Already in 1746 was it stated by one researcher, P. Fauchard, that the passions may be reckoned as internal causes which produce disease in the teeth. In 1879, another researcher claimed that the overwrought brains of 'our civilized children' may be a factor in the causation of dental decay. There have been other references to similar suggestions at the end of the last century and early in the present one. In 1938, an editorial in the *Dental Digest* claimed that the prognosis of dental caries was partly conditioned by mental states.

This observation doesn't hold all the way. People living under wartime conditions and especially prisoners of war and inmates in concentration camps generally have a low caries rate. But it is also known that wherever there is starvation, the caries index is low and any caries will be of the arrested type. It is possible, however, under normal conditions, that changes in the blood flow to the pulp as a result of mental stress may cause some as yet unidentified change to take place in the teeth.

All this is perhaps of little use to those who want to avoid visits to the dentist, but it shows how much work still needs to be done, and how little the public has been told. It also shows up just how naïve the advice the public *is* given, can be.

It is remarkable how the toothbrush-toothpaste myth is invariably found exactly in those places where tooth decay is most rampant. This modern gimmick is totally linked up with the dental plaque and acid philosophy. The hard modern brush is supposed to scrape the plaque off the flat surfaces of the teeth – usually demonstrated on the surface of the front teeth where caries never occurs – while it is utterly useless for poking out the muck *between* the teeth. Dental heretics have claimed for decades that only soft brushes should be used, if at all. Most recently Dr Merril K. Wheatcroft at the University of Texas told the Annual Greater New York Dental Meeting that hard bristles may puncture and lacerate the gums, causing them to recede, and expose roots.

As for the pastes, some of the best-known US makes were indicated by the FDA in 1970 for the false advertising claims that they reduce decay. Ralph Nader had a go at them in 1971. From time to time even perfectly orthodox dentists come out and explain how toothpastes merely smear the teeth and cause food debris to get retained in the mouth.

The Danish dental researcher, H. Eggers Lura, says:

'The recommenders of toothpastes have never tried to make control experiments of mouth rinsing purely with cold, oxygenated water, in contrast to the dirty, greasy glycerin- and sugar-containing paste ingredients. There are several examples where these toothpastes have had a caries-promoting effect. Most toothpastes contain insoluble polishing ingredients which are able to bind sugar and retain it in the mouth unless thoroughly rinsed away. sticky toothpastes soil the mouth – they only cover the bad smell from food retentions.'

Another odd aspect of modern pastes is the way in which they are often given a minty taste. This even goes for some toothpicks. And yet, clinically, peppermint is caimed to be the most devastating of all cariogenic substances, as many ex-smokers have discovered to their grief when switching to sucking peppermint instead of cigarettes. People have simply come to associate a minty taste with a 'clean' mouth. Even those dentists whose research is financed by the toothpaste industry have from time to time felt compelled to fight against this one. Yet we have become a generation who find it almost impossible to give up the toothpaste habit. This habit *does* hide the results of the sort of food we eat, at least temporarily and overnight, and stops us thinking about what really ought to be done about our teeth. Anyone who cares to, can perform a small 'scientific' experiment in his own bathroom, trying to clean his teeth with soft toothpicks and lots of water and timing with a stopwatch just how long it takes to get rid not only of the food debris but the bad smell – and he will have some idea of the magnitude of the problem.

Just how vicious some commercial dental feuds may get has been illustrated by the saga of the Scandinavian row over Bofors, the huge industrial concern in Sweden.

In late 1971 and early 1972, Bofors came out with a non-abrasive toothpaste. Not an ideal way, perhaps, to woo a profession who go in for such relatively brutal methods as scaling the teeth, but it was a brave attempt. That Christmas, full-page advertisements appeared in the national newspapers in which Bofors stated what had previously only been whispered in dental laboratories and discussed with journalists only after elaborate promises of confidentiality and threats that everything would be denied if quoted.

With the backing apparently from researchers at Lund University in Sweden and the Academy of Medicine in New Jersey, Bofors told the world that the polishing and scaling ingredients in 'other' toothpastes might damage the teeth. The new Bofors toothpaste, they claimed, would cut or prevent such damage by 50 percent. The ads showed magnified photographs of the ingredients in the 'other' pastes which, the text said, consisted of chalk with sharp edges and corners that could grate down the enamel and the dentine near the gums. The Bofors paste consisted of minute plastic balls which did not damage the teeth.

Not surprisingly, the manufacturers of the 'other' pastes – most of the leading names – as well as a number of dental academics, descended on Bofors like a pack of wolves. The following day the leading newspaper in Denmark, *Politiken*, carried a lengthy article with a tape-recorded discussion between some of the leading players in the drama. At the same time one of the 'other' manufacturers came out with an advertisement recommending their own paste on the grounds that it contained a quartz with hardness 7 which, they said, could be relied upon to really scour the teeth. The results? Total and utter confusion among the Danish public. Some of those I talked to even worried about whether their *false* teeth might be damaged by ordinary products.

Later the following year, the controversy was complicated by a claim, happily supported by the Bofors opponents, that the little plastic balls would pollute the environment and cause widespread cancer. Subsequently a further complication arose when the Swedish and Norwegian anti-fluoride groups and environmentalists came out against cancer and little plastic balls – practically holding hands with Professor Yngve Ericsson who had been quick to defend his own highly abrasive toothpastes.

By then the whole thing had become ludicrous. As an exercise in how to confuse and mislead the public the whole epidose was superb. It is not easy to decide which of the commercial interests involved were the most wicked, but some of the comments from the dental experts reported in *Politiken* were very enlightened. Thus Alis Moss, of the Copenhagen Dental College, when commenting on the dental plaque which the scouring is supposed to remove, said, 'It is very difficult to decide how much should be removed . . . and how much should be left behind in order to *protect the teeth*'.

There were other academics who wanted not only the Bofors ads, but the paste as well, to be withdrawn and who produced instant tests with results that apparently proved that the Bofors paste stained the teeth. The Bofors scientists suggested that the 'other' toothpaste companies should also be asked to produce *scientific* evidence, despite the instant evidence, and hinted mysteriously, *'They haven't got the courage . . .'*.

And this, perhaps, sums it all up. Dentistry takes a lot of courage. Why this one field of medicine should have got itself into this state is not easy to see, unless it is because the public themselves have come to regard dental health and teeth as a kind of cosmetic. There are many similarities between the dental market and the make-up business and, to some extent, the public must also bear part of the blame.

Today some 17 million Britains are toothless. One researcher has calculated that children's teeth are decaying at the rate of one every two and a half seconds. Children of two and a half are queueing up for dentures. The manufacturers of preventive gadgets pick up £23 million a year, and chemists' magazines and trade journals talk optimistically of the tremendous 'growth potential':

'US dentists are now doing 28 million prescriptions a year and the average dentist spends 3,000 dollars a year himself on dental supplies including drugs. Still dental disease is the most widespread chronic ailment in the US'. And in Britain too.

The remedy? More disclosing tablets to show up the presence of plaque, more dental floss, more electric appliances for home use, more toothpastes, more toothbrushes, more magnifying mirrors with high-intensity lamps to give the patient the best view of the damage, more home ultrasonics for removing tartar and calculus deposits. If all that fails, more aerosol cleaners for false teeth. Total denture adhesive sales in the US were expected to reach 44 million dollars in 1973.

And no wonder. In 1972, Herbert S. Denenberg, insurance commissioner for the Commonwealth of Pennsylvania, claimed that six million teeth were extracted each year in the US unnecessarily. He also claimed in his report – *A Shopper's Guide to Dentistry* – that 'there is evidence that much dental care is substandard' and 'dentists have also been slow to introduce accepted methods of dental disease prevention'.

They have certainly been slow to introduce *accepted* methods. They have introduced *effective* methods hardly at all. They can repair rotting teeth brilliantly. But there is nowhere in Britain today where you could go and get good medical advice about how to prevent dental disease (except perhaps from paediatricians).

By the year 2000, it has been predicted by some, the nation may be practically toothless from the cradle.

'When the new preventionists,' says Robert S. Feller, a dentist in Thousands Oak, California, criticizing his colleagues for wasting time on fancy-sounding gimmicks in the *Dental Survey* in August 1972 – 'when the new preventionists talk about their great ideas, they are milling over again old products . . . neglecting the many other aspects of prevention, diet, habit patterns, preventive orthodontics, heredity, gingival massage and elimination of pre-existing pathology . . . it is about time that dentistry stopped being so naïve and jumping on a bandwagon and started remembering that they have in their hands a fine and necessary service to do. *There isn't any magic.*'

There have, as a matter of fact, been one or two pieces of magic. As a semi-preventive measure the fissure sealants deserve to be mentioned. A new British improvement on the early American sealants has just appeared, and this is one way in which caries can actually be 'prevented'.

I would still describe it as, at best, a stop-gap measure, in every sense of the word. It is still a question of treating symptoms and not the cause of the disease.

A child who has been bottle-fed, who has received a somewhat second-rate nutrition during infancy, and who eats sweets and unhealthy food throughout childhood, may have his teeth 'lacquered' and apparently avoid cavities as long as the treatment is repeated at regular intervals. But he will continue to consume the wrong sort of food and this would sooner or later affect his general state of health, including his mouth. It is my own guess that when the present generation of fissure-sealed children reach early adulthood, a new set of problems may well present themselves. Certainly gum disease would be as rampant as before and thus the patients would lose their teeth prematurely anyway.

I have tried to put these questions to the fissure-sealant experts and met with a blank refusal to answer or discuss my points. Probably the truth is that nobody knows what will happen in twenty years' time to fissure-sealed teeth. It remains a stop-gap measure, though a good one. But it should properly be regarded as a surgical and clinical procedure and not as a 'preventive' one. It is there to cope with a situation that has already gone wrong.

It is a sad, though splendid, example of the sheer amount of brilliant scientific effort and knowledge we put into mending and fighting problems which needn't happen in the first place, and which we already know how to prevent.

Fluoride terminology

Since the 1940s fluoride promoters have developed a vocabulary all their own. It is necessary to know some of these to understand the debate at all.

mineral nutrient fluoride: invented by Frederick Stare, of Harvard, and refuted by the FDA and the US Public Health Service.

fluoride deficient teeth: if taken to mean that the teeth lack something they must have, or have been deprived of, this term is grossly misleading.

optimum level: originally meant the optimum *permissible* water level as a guide to polluters. Has now come to mean the optimum desirable level, as a guide to dentists. Is undesirable either way.

demonstration: since 1951 it has been the custom to talk of 'demonstrations' and not 'experiments' when discussing *human* consumption of fluoride, since it was thought that people wouldn't like being 'experimented' on.

controlled: Fluoridators have an almost fanatical obsession with the word 'controlled'. They do not use the word in the normal scientific sense but merely to indicate that the amount of fluoride added to the water is controlled. Little else is, least of all the amounts actually consumed by peope.

benefit: fluoride promoters find it difficult to talk about fluoride at all without linking it with the word 'benefit' as noun, verb or adjective.

protection: as with the word 'benefit'. Thus a sentence may read: 'a demonstration of controlled fluoridation at the optimum level proved the benefit of protection of fluoride'. The antis in turn talk about a '*protection racket*'.

artificial: this word is abhorred by all promoters and is essentially a word used by the antis whenever the pros talk about 'controlled' fluoridation.

natural: used as the ultimate proof of the safety of controlled fluoridation. Is based on the idea that 'nature set this thing up'. When all other arguments fail, promoters will resort to calling on nature to prove their case.

cuts dental decay by half: a slogan which makes other scientists laugh. Is a reduction percentage: 1 tooth is 50 percent of two, two are 50 percent of four, 8 are 50 percent of 16. Says nothing about the actual number of teeth involved, and in Britain means 0.9 of a tooth's difference.

mass medication: whatever fluoride is, the promoters insist that it is not mass medication, and it is not medication. Nature set this thing up and it's a mineral nutrient.

vociferous minority: a group which includes all antis, from Barry Commoner to the Ku Klux Klan and the John Birch Society.

communist plot: fluoride promoters claim that the antis claim that fluoridation is a communist plot.

un-American: (in the US only): the vociferous minority.

rat poison: fluoride promoters claim that the antis always describe fluoride as rat poison. Some of them do. The word is therefore used frequently by the pros to discredit the antis.

hoax, plot, conspiracy: words used equally often by both sides.

psychopath, fanatics, hate-mongers: used by the pros about the antis.

dental mafia: used by the antis about the pros.

nutcase: used by some doctors about the pros.

absolutely: fluoride is absolutely safe!

mis-informed: anyone who doesn't believe that.

List of further reading

Environmental Fluoride, by J. R. Marier and Dyson Rose, Environmental Secretariat, Division of Biology, National Research Council of Canada, Ottawa, Canada. NRC Publication No. 12,226. 1971.

Controlling the Potential Hazards of Government-Sponsored Technology, by Michael Wollan, The George Washington Law Review, Volume 36, Number 5, Page 1105, July 1968.

Two Issues of Science and Public Policy: Air Pollution Control in the San Francisco Bay Area and Fluoridation of Community Water Supplies, (Section on fluoridation pp. 148-489, Introduction pp. 1-9), by Dr Edward Groth III, Department of Biological Sciences, Stanford University, 1973. Available from University Microfilms, 300 North Zeeb Road, Ann Arbor, Michigan 48106. (1973).

Fluoridation: Errors and Omissions in Experimental Trials, 2nd Edition, 1960, by P. R. N. Sutton, Melbourne University Press.

Health Effects of Environmental Pollutants, by George L. Waldbott, The C.V. Mosby Company, Saint Louis, 1973 (distributed in Britain by Henry Kimpton, Publishers).

Fluorides and Human Health, by Michael J. Prival, Center for Science in the Public Interest, 1779 Church Street, N.W., Washington D.C. 20036. 1972.

Fluorides in the Air, by Michael J. Prival and Farley Fisher, Center for Science in the Public Interest, 1972.

Index